COME RAIN OR SHINE

OR SHINE

A WEATHER MISCELLANY

STORM DUNLOP

summersdale

COME RAIN OR SHINE

With research by Katherine Bassford

Illustrations © Shutterstock

Summersdale Publishers Ltd
46 West Street
Chichester
West Sussex
PO19 1RP
UK

www.summersdale.com

Printed and bound in the Czech Republic

ISBN: 978-1-84953-718-6

Substantial discounts on bulk quantities of Summersdale books are available to corporations, professional associations and other organisations. For details contact Nicky Douglas by telephone: +44 (0) 1243 756902, fax: +44 (0) 1243 786300 or email: nicky@summersdale.com.

CONTENTS

INTRODUCTION

One of the advantages of living in Britain is that we experience very changeable weather. This not only serves us as a useful topic of conversation when talking to strangers, but does actually reveal that, over the years, Britain has experienced some remarkable weather-related events. This book aims to touch on some of these and, with its sections of facts and figures, show some of the great range of conditions that have occurred in the past. Making the choice of which events, people and other factors to describe has been very difficult, because there are so many from which to choose. In this respect, Britain is particularly favoured, because reliable British weather records exist for a longer period than for any other country in the world. Unfortunately many people tend to think of weather purely in terms of weather lore, which is usually unreliable although a few sayings do contain a grain of truth and have therefore been included here. British weather has produced some surprising and fascinating events, many of which are covered in the following pages.

WEATHER LORE

Although there is a vast amount of weather lore – there are whole books devoted to it – unfortunately most sayings prove, on closer examination, to be invalid. Even in a relatively small country such as Britain, the weather is rarely the same in all parts of it, so many sayings are relevant to a small area only, for a very short period of time, or not every year. One of the most well-known sayings, for example, is that if it rains on St Swithin's Day, 15 July, it will rain for a total of 40 days. Unfortunately, study of weather records shows that this is simply not true and in any case, it would certainly not be true for the whole of Britain. But, worse: the same prediction is also made of rain on St Medard's Day, 8 June, and St Protasius's Day, 19 June. Given that St Swithin's Day falls within both of those 40-day periods, the number of rainy days would be even greater. It is, in fact, perfectly possible to find similar sayings about a whole list of saints' days and construct a sequence, showing that it 'must' rain on every single day of the year. The beginning and end of such a sequence overlap, so once it started to rain, it would never stop!

Many of the sayings of weather lore are contradictory. There is, for example, a well-known rhyme about trees coming into leaf and rain:

'If the ash before the oak,
We shall have a soak.
But if the oak before the ash,
We shall have just a splash.'

Unfortunately, there is a similar rhyme that predicts the exact opposite:

'If the oak is before the ash,
'Twill be a summer of wet and splash.
If the ash is before the oak,
'Twill be a summer of fire and smoke.'

USEFUL NOTES

Greenwich Mean Time

All meteorological observations are taken at specific, standardised times of the day. Following the guidelines established by the World Meteorological Organization, observations are made and reported worldwide using what is known as Coordinated Universal Time (UTC). For all intents and purposes this may be considered to be the same as Greenwich Mean Time (GMT). Observers around the world take their readings at the standard times, regardless of the local time zones in which they are located. This obviously enables data to be compared worldwide and is essential for forecasting purposes. In Britain, a standard meteorological day runs for 24 hours from 09.00 GMT on one day to 09.00 the next day. (Summer Time is not used.) Confusion sometimes arises when 24-hour extreme values are quoted. A standard rain day, for example, covers the 24 hours from 09.00 GMT, whereas figures quoted for extreme rainfall in 24 hours may begin and end at any time.

In the case of British rainfall, to continue with the example, the greatest rainfall in one rain day occurred at Martinstown in Dorset on 18 July 1955 with 279.4 mm

(11 in) of rain. The most extreme rainfall in 24 hours occurred at Seathwaite in Cumbria from midnight to midnight on 19–20 November 2009, when 316 mm (12.44 in) of rain was recorded.

Temperature

It is easy to become confused between actual temperatures and changes in temperature. You cannot use a single conversion table otherwise dreadful errors creep in. Meteorologists get over the problem by always showing actual temperatures as '°C' – such as '17°C' – and changes (or differences) as 'degrees (or deg.) C', e.g. 'Rising air cools at about 10 deg.C per kilometre.'

Air pressure

Most people in Britain are now used to air pressures given in millibars. In a very few places (such as the United States) pressures are still sometimes given in the old units of inches of mercury (in/Hg), describing the length of a mercury column in an old-fashioned barometer. Meteorologists still sometimes use millibars (mb), but increasingly give pressures in hectopascals (hPa), based on the standard, Système Internationale (SI) unit the Pascal. However, one millibar (1 mb) is absolutely identical to one hectopascal (1 hPa), so no conversions are required.

Gregorian calendar

It should be noted that a few of the events mentioned in this book occurred before the major calendar reform by Pope Gregory (in 1582) was implemented in Great Britain. This change was from what is known as the Julian calendar to the Gregorian calendar. Although the change was adopted rapidly in Catholic countries, many Protestant countries changed decades or even centuries later. In Britain (and its colonies at the time) the change to the Gregorian calendar did not take place until as late as 1752. The dates in the original Julian calendar are known as 'Old Style' (OS) and in the Gregorian as 'New Style' (NS). Dates of some British events before 1752 are given in both forms, so that the New Style dates may be directly compared with the modern calendar. The Great Storm of 1703, for example, occurred on 26 November 1703 (OS) or 7 December 1703 (NS).

CHAPTER 1

SUNSHINE

A cloudy day or a little sunshine have as great an influence on many constitutions as the most recent blessings or misfortunes.

Joseph Addison

Britain may have a reputation (especially among people living in Continental Europe) for dreary, dismal, rainy or even foggy weather, but it does, in fact, enjoy a considerable amount of sunshine. This is borne out by the fact that consistent British weather records cover a much longer period of time than those for any other region of the world, and those records reveal that Britain does enjoy many periods of sunshine and even prolonged droughts.

WEATHER PEOPLE

Gordon Valentine Manley (1902–1980)

'Central England Temperature' is a term known to all meteorologists. It is a record of the mean monthly temperatures in the centre of England covering the whole period from 1659 to date and it is indissolubly linked with the name of Gordon Manley.

Manley was born in Douglas, Isle of Man, but grew up in Lancashire. He made meteorological observations whilst still a young boy. He read engineering at Manchester University, graduating in 1921, then read geography at Caius College, Cambridge, graduating from there in 1923. He was employed by the Meteorological Office at Kew Observatory for a short period, 1925–26, but then took up a lectureship at Birmingham University. He moved to Durham University in 1928. There he began his work on British climate records for which he was to become famous. He established an observatory in the northern Pennines and carried out observations (most particularly of the Helm wind) under extremely arduous conditions.

He moved back to Cambridge in 1939 and remained there until 1948 when he accepted the chair of geography at the University of London's Bedford College. By this time he had become interested in research into climate change (in which he was a pioneer) and palaeoclimatology. He was heavily involved with various societies and committees, including being part of the British committee for the International Geophysical Year (IGY), which ran from 1 July 1957 to 31 December 1958. It was in 1953 that he first published his work on historic temperature records for central England for the period 1698–1952, which he expanded in 1974 to cover the whole period from 1659 to 1973. This Central England Temperature (CET) series of monthly mean temperatures is the longest standardised instrumental record available for anywhere in the world. It has subsequently been kept up to date by the Meteorological Office (now known as the Met Office).

> Good weather is like good women –
> it doesn't always happen and when
> it does it doesn't always last.
>
> Charles Bukowski, 'Cows in Art Class'

NOTABLE WEATHER EVENTS

The Hot Spell, 1990

Although the summer of 1976 is notorious for the prolonged drought, and the introduction of a 'drought minister', who no sooner started work than it began to rain, Britain has experienced many other periods with little rain and high temperatures. One such period occurred in early August 1990. A new British temperature record of 37.1°C was set at Cheltenham on 3 August, beating the previous figure of 36.7°C set at Raunds in Northamptonshire, Epsom in Surrey and Canterbury in Kent in 1911. On that day, temperatures of 35°C were recorded over a wide area of eastern Wales, southern, central and eastern England. Although relatively short in duration (1–4 August), the area over which the high temperatures prevailed was much greater than for any other period in the twentieth century.

The extreme heat produced many problems for transport. Many road surfaces and one runway at Heathrow Airport melted in the heat, and the railways also suffered, with rails buckling in many locations as they expanded with the high temperature. This led to widespread speed reductions, especially of the high-speed intercity trains. There were numerous farm and heathland fires, and on both 3 and 4 August there were several deaths from drowning, where people ill-advisedly plunged into cold water to escape the high temperatures. One slightly bizarre incident occurred in Liverpool, where the entire stock of a chocolate factory melted in the heat.

Britain's hottest day, 2003

Sunday, 10 August 2003 saw temperature records broken over a large area of Britain and particularly in southern England. The day was warm in Northern Ireland and

Scotland, but in England it proved to be extremely hot. A new United Kingdom temperature record of 38.5°C was set at Brogdale, near Faversham in Kent, and long-term station temperatures were broken at many stations in the London area and the Home Counties. Odiham in Hampshire recorded 13.1 hours of sunshine.

As a result of the heat a band of thundery rain developed over North Wales, the West Country and the western Midlands shortly after dawn. This moved eastwards, gradually weakening, although at Carlton-in-Cleveland in North Yorkshire there was 48 mm (1.89 in) of rain in 15 minutes and 2 cm (0.79 in) of hail at around 09.50 GMT.

The Big Drought, 2010–12

Surprising though it may seem, Britain does occasionally suffer from prolonged periods of drought. There was below-average rainfall for a series of months, beginning in the winter of 2009–10 and continuing until March 2012. The spring, autumn and winter months were particularly affected. The only comparable two-year period of drought in the past 100 years occurred between April 1995 and March 1997.

There was an extended dry spell from January to June 2010, and this resulted in water shortages in north-western England. Wet weather there in July and then over the south-east in August eased the situation. However, in the next year, 2011, spring proved to be extremely dry, particularly in the east of England, although river and groundwater levels were low throughout the country. There were widespread wild fires in Northern Ireland, in the Highlands of Scotland, in mid-Wales, Lancashire and Berkshire.

A declaration of drought conditions was made in June 2011, covering central and eastern England and, despite a poor summer, the autumn remained dry, giving difficult conditions for the harvesting of crops. The following winter also remained dry, and in March 2012 the area covered by drought orders was extended to cover most of southern, central and south-eastern England. More wild fires occurred in Surrey, south Wales and in the Scottish Borders. It was only exceptionally heavy rainfall in the months of April to July 2012 that brought an end to this extended period of drought.

MYTHS AND MISTAKES

The 'barbecue summer' that wasn't

At the end of April 2009, the media were full of reports that the Met Office was predicting a 'barbecue summer', with sizzling hot temperatures after the dreadful summers of 2007 and 2008. But the Met Office didn't predict high temperatures. It made a mistake in its presentation to the media, and showed an image with a red-tinted area, covering most of the country, where temperatures would be slightly higher than normal. The media took that to mean that most of the country would be 'extremely hot', whereas in fact, the temperature was expected to be just 1.5 degrees C warmer than normal. In reality, that prediction was more or less correct, but there were accompanying high winds and heavy rainfall. One might say that instead of being particularly hot, the summer of 2009 was 'rained out'.

An 'Ark' in the sky

In his light-hearted book, *The Weather Eye*, C. R. Benstead, a forecaster, describes how a Meteorological Office forecast of a serious deterioration in the weather was decried in a local paper by a farmer who proclaimed, in effect, that 'Not so! I have seen an Ark in the sky, pointing in the right direction, and not only will we have a fine weekend but the weather will continue fine for several days to come.' The weather broke on the Thursday. Benstead attended a discussion where the senior forecaster – a person whose pronouncements were regarded as gospel – described the general situation. At the end of the discussion, Benstead asked 'Can you tell me what an Ark in the sky is?' No one deigned to give him an answer.

In fact, the farmer was doubtless wrong: it was not a Noah-type 'Ark', but an 'arc', almost certainly a portion of a solar halo, a circle around the Sun, caused by light from the Sun being refracted by ice crystals, high in the atmosphere. Rather than being a sign of fine weather, such a solar halo is instead one indication that the weather will deteriorate. The ice crystals are usually in the form of a thin sheet of cirrostratus cloud, often so thin that it is undetectable and its presence only revealed by the halo. Such cirrus and cirrostratus cloud is a typical forerunner of a warm front. Wisps of cirrus gradually thicken into a sheet of cirrostratus, which itself thickens into lower, altostratus clouds, when the Sun gradually becomes obscured and finally appears as if seen through ground glass and fails to cast any distinct shadows. As the warm front continues to approach, the sheet of altostratus itself thickens and eventually becomes nimbostratus, covering the whole sky in a dark grey cloud from which rain begins to fall.

Solar haloes are common. In Britain, they may be seen, on average, about once every three days, such is the frequency of fronts that cross the country.

FASCINATING FACTS:
THE PURPLE LIGHT

On very rare occasions, the sky at sunset or sunrise, becomes filled with a vibrant, purple light, completely different from the usual tints of yellow, orange, pink or red. The purple shade is highly distinctive and usually very striking. It only occurs when powerful volcanic eruptions have ejected material high into the atmosphere, particularly large amounts of the gas sulphur dioxide. This combines with water to actually form fine droplets of sulphuric acid, which spread around the globe, carried by winds in the stratosphere. The normal colour of the sky is blue, of course, where blue light is scattered by the molecules of oxygen and nitrogen in the air. Red light is not scattered by those molecules, which is why red colours appear at sunrise and sunset, when all, or nearly all, of the blue light has been scattered aside during the light's long path through the atmosphere, so only the yellow to red tints reach the observer.

However, the tiny sulphuric-acid droplets, suspended in the atmosphere, do scatter red light, which mixes with the normal blue light to give the highly distinctive purple shade.

But because the combination of large volcanic eruptions and suitable weather patterns to spread the gases around the Earth is rare, the purple light appears on very few occasions. The last major eruption that produced purple skies over Britain was that of Mount Pinatubo in the Philippines in 1991. The previous occasion was years before, after the eruption of El Chichón in Mexico in 1982. Since then there have been no major eruptions and suitable weather conditions to spread material over Europe.

Sometimes the ash that volcanoes eject high into the atmosphere will spread around the Earth, but this gives rise to completely different colours in the sky, rather than the purple tint. In the case of ash, the high layer tends to appear in an orange tint and often shows a fibrous or striated appearance, where the bands run across the sky, roughly at right angles to the light from the rising or setting Sun.

WEATHER LORE

Even the well-known saying about the 'red sky at night' has to be interpreted with care. It is, of course, very ancient, going back to the Bible:

> 'When it is evening, ye say,
> It will be fair weather: for the sky is red.
> And in the morning,
> It will be foul weather to day: for
> the sky is red and lowering.'

Matthew 16:2–3

More commonly, nowadays, the rhyme runs something like this:

> 'Red sky at night, shepherd's delight,
> Red sky at morning, shepherd's warning.'

(Sometimes it is 'sailor's delight' and 'sailor's warning', or some similar wording.) There is a certain sense in which this is correct, in that, because most weather systems move from west to east, if the sky is clear to the west, red light from the setting Sun will illuminate the clouds with a red glow, and any poor weather may well pass. Whereas if the clouds of an approaching depression are encroaching on the sky from the west, light from the rising Sun will illuminate them with red, but the clouds will increase and the weather deteriorate. Even so, the exact shade of red does have implications, because it is affected by the amount of moisture in the air. A pale red or pinkish tint to the Sun itself at sunset or sunrise does suggest fair weather for the next few hours, but a deep 'angry' red ('lowering' in the words of the Bible) tends to indicate that wind and rain may well follow.

> 'So it falls that all men are
> With fine weather happier far.'

John Fitchett, *King Alfred*

HIGHS AND LOWS

Highest daily maximum temperature by country

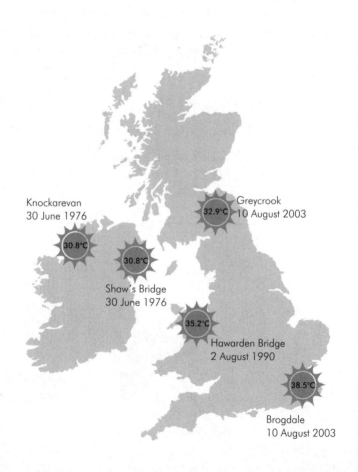

Knockarevan
30 June 1976

30.8°C

30.8°C

Shaw's Bridge
30 June 1976

Greycrook
32.9°C 10 August 2003

35.2°C

Hawarden Bridge
2 August 1990

38.5°C

Brogdale
10 August 2003

Lowest daily minimum temperature (United Kingdom record)

The record for the lowest temperature observed in the UK, –27.2°C, is shared by two locations, being recorded twice at Braemar, Aberdeenshire, on 11 February 1895 and 10 January 1982, and on 30 December 1995 at Altnaharra, Highland.

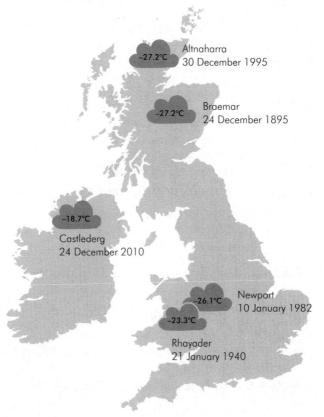

Highest minimum daily temperatures by country

(In other words, the temperature that day never dropped below the figure given.)

England: 23.9°C on 3 August 1990 at Brighton, East Sussex

Northern Ireland: 20.6°C on 31 July 1868 at Armagh, County Armagh

Scotland: 20.5°C on 2 August 1995 at Creebridge, Wigtownshire

Wales: 22.2°C on 29 July 1948 at Victoria Park, Swansea, West Glamorgan

Lowest daily maximum temperatures by country

(In other words, the temperature never rose above the figure given.)

England: −11.3°C on 11 January 1982 at Newport, Shropshire

Northern Ireland: −11.3°C on 23 December 2010 at Edenfel, County Tyrone

Scotland: −15.9°C on 29 December 1995 at Fyvie Castle, Aberdeenshire

Wales: −8°C on 12 January 1987 at Trecastle, Dyfed

Sunniest month

The month with the most hours of sunshine was July 1911, when Eastbourne in Sussex (now East Sussex), recorded 383.9 hours of clear skies.

Sunniest place

The location that records more sunshine than any other in the British Isles is Bognor Regis in West Sussex, with an average of 1,902.9 hours of sunshine per year.

Dullest place

The location with the least number of hours of sunshine in a year proves to be the summit of the mountain of Ben Nevis, near Fort William in the Scottish Highlands, with a yearly average of just 736 hours.

Highest monthly sunshine records by country

England: 383.9 hours in July 1911 at Eastbourne, Sussex

Northern Ireland: 298 hours in June 1940 at Mount Steward, County Down

Scotland: 329.1 hours in May 1975 at Tiree, Argyll and Bute

Wales: 354.3 hours in July 1955 at Dale Fort, Pembrokeshire

Highest recorded pressure

The highest pressure recorded in the British Isles is 1,053.6 millibars, observed at Aberdeen Observatory in Aberdeenshire, on 31 January 1902.

Lowest recorded pressure

The lowest pressure (916 millibars) was recorded at North Rona during the Braer Storm (see page 130).

A change in the weather is sufficient to recreate the world and ourselves.

Marcel Proust

HIGHEST DAILY MAXIMUM TEMPERATURE RECORDS (UK)

Month	Temp (°C)	Date	Location
Jan	18.3	10 January 1971	Aber (Gwynedd)
		27 January 1958	Aber (Gwynedd)
		26 January 2003	Aboyne (Aberdeenshire)
		26 January 2003	Inchmarlo (Kincardineshire)
Feb	19.7	13 February 1998	Greenwich Observatory (London)
March	25.6	29 March 1968	Mepal (Cambridgeshire)
April	29.4	16 April 1949	Camden Square (London)
May	32.8	22 May 1922	Camden Square (London)
		29 May 1944	Horsham (West Sussex)
		29 May 1944	Tunbridge Wells (Kent)
		29 May 1944	Regent's Park (London)

Month	Temp (°C)	Date	Location
June	35.6	28 June 1976	Southampton, Mayflower Park, Hampshire
July	36.5	19 July 2006	Wisley (Surrey)
Aug	38.5	10 August 2003	Faversham (Kent)
Sept	35.6	2 September 1906	Bawtry, Hesley Hall (South Yorkshire)
Oct	29.9	1 October 2011	Gravesend (Kent)
Nov	21.7	4 November 1946	Prestatyn (Denbighshire)
Dec	18.3	2 December 1948	Achnashellach (Highland)

If the cows lay down, it meant rain. If they were standing it would probably be fine.

Terry Pratchett and Neil Gaiman, *Good Omens: The Nice and Accurate Prophecies of Agnes Nutter, Witch*

WEATHER WORDS

Dealramh na gréine (Irish): Sunshine.

Enfys y bora, aml gawoda. Enfys y p'nawn, tegwch a gawn (Old Welsh): Morning rainbow may bring showers. Afternoon rainbow fine weather we'll have.

Fox (Pembrokeshire): A single fine day between spells of bad weather.

Glassy Sun (Pembrokeshire): A bright Sun with no hint of colour is taken as a sign of forthcoming rain. This probably refers to the Sun shining through altostratus clouds, when it appears as if seen through ground glass and casts no shadows. With an approaching warm front, altostratus generally thickens into heavy, grey nimbostratus cloud that produces heavy rain.

Griain (Irish): Sun.

Haul dan ei gaerau (Old Welsh): Sunset; literally, the Sun going under its ramparts.

Haul y gwanwyn, gwaeth na gwenwyn (Old Welsh): Spring sunshine worse than poison.

Leesome (Scots): Fine.

Mae'n heulog (Welsh): It's sunny.

Mochy (Scots): Close and, in particular, extremely humid.

Ni adawodd haf sych erioed newyn ar ei ol (Old Welsh): A dry summer never resulted in a famine.

Simmer cowt (Scots): An unusual term for a heat haze.

Slatch (Sussex dialect): A brief respite or interval in the weather.

Swallocky (Sussex dialect): A term used for sultry, humid and unpleasant weather.

Te (Irish): Hot.

Tirim (Irish): Dry.

Wherever you go, no matter what the weather, always bring your own sunshine.

Anthony J. D'Angelo

CHAPTER 2
RAIN

After three days men grow weary, of a wench, a guest, and weather rainy.

Benjamin Franklin

WEATHER PEOPLE

Sir Gilbert Thomas Walker (1868–1958): Monsoons and teleconnections

Nowadays, practically everyone has heard of El Niño. To meteorologists the phenomenon is known as ENSO – the El Niño-Southern Oscillation. The 'Southern Oscillation' is part of the 'Walker circulation', named after Sir Gilbert Walker. Although not the first to be discovered, this brought the concept of teleconnections – links between atmospheric or oceanic conditions in widely different locations – to the wider attention of the meteorological community.

Walker was a mathematician and meteorologist. Born in Rochdale in Lancashire, he eventually obtained a scholarship to Trinity College, Cambridge. There, he excelled in mathematics, being senior wrangler in the examinations in 1889 and 1890, and became a fellow of Trinity in 1891. His initial studies were of electromagnetism, but he then became specifically interested in the physics of projectiles and was particularly expert on all primitive forms of projectiles, such as boomerangs, earning the nickname Boomerang Walker.

Walker moved to India in 1904 and became director of observatories. He was particularly involved in the study of the monsoon – a vital matter in India, where agriculture is dependent on the monsoon rains. (There was a major famine in India, caused by widespread crop failure, in 1899–1900.) He attempted to develop methods of predicting the extent of the variations in monsoon rains, but this problem proved insoluble. In the course of these studies he found a correlation between pressures at widely separated locations over the Indian and Pacific oceans. For example, when pressure rises over Tahiti, it falls over Darwin in northern Australia, and vice versa. This phenomenon is now known as the Southern Oscillation. Subsequent work has shown the existence of a number of similar relationships, now known by the general term 'teleconnections'. Walker also showed a relationship between rainfall over India and Java and pressures in the Pacific, and established the importance of large-scale, zonal circulations in the tropics, rather than the meridional circulations that drive most of the processes in temperature zones. (Zonal circulation is generally along lines of latitude, rather than in a north–south direction.) Such zonal circulation cells are known as Walker cells, and the overall motions as a Walker circulation.

His observation of soaring and gliding Himalayan birds led him to the study of flight and he later became an expert in the physics of human gliding and soaring flight, doing much to encourage the adoption of the hobby in Britain.

> One can find so many pains
> when the rain is falling.
>
> John Steinbeck

NOTABLE WEATHER EVENTS

The Lynmouth Disaster, 1952

On 15 August 1952, torrential rain fell on Exmoor in north Devon, where the ground was already saturated by earlier rains. At the Longstone Barrow weather station on Exmoor, 290 mm (11.42 in) of rain were recorded in 24 hours, and the maximum rate between 20.30 and 22.30 GMT has been estimated at more than

25 mm (0.98 in) per hour. The water poured off the northern side of Exmoor and a dam of debris formed on the West Lyn river. This then collapsed, sending a surge of water, rocks, tree trunks and other debris down the extremely steep watercourse, and into the town of Lynmouth, destroying 28 of the 31 bridges, and more than 100 buildings. A total of 34 people were killed, and serious damage was done to the seawall and lighthouse (the latter collapsed the next day). Dozens of cars were washed out to sea and some were never located.

Rain scape

The Eskdalemuir Storm, 1953

On 26 June 1953, Eskdalemuir in Dumfriesshire experienced torrential rain, producing localised flooding and some damage. In 24 hours, 106 mm (4.17 in) was recorded, with as much as 80 mm (3.15 in) in one 30-minute period. There was more heavy rainfall farther south, with more extreme rainfall rates. At Holehird,

Windermere, in Westmoreland, 72.1 mm (2.84 in) fell in 55 minutes; 53.1 mm (2.09 in) in 39 minutes at Langley in Cheshire; and 43.7 mm (1.72 in) in 15 minutes at Nelson in Lancashire. By contrast, in southern England the weather was very warm, and locally even hot, with 14.3 hours of sunshine at Lympne in Kent.

The Martinstown rainfall event, 1955

The record for British rainfall during an official observational day was gained by Martinstown, near Dorchester, in Dorset, on 18 July 1955, where 279 mm (10.98 in) of rain fell in a single day (09.00 to 09.00 GMT, 18–19 July). Most of the rain fell during a 15-hour period. Martinstown – formally known as Winterbourne St Martin, where 'Winterbourne' means that the small river flows mainly during the winter – suffered extensive flooding some hours after the peak downpour. Later analysis by the Meteorological Office suggested that even heavier, unrecorded, rainfalls of over 305 mm (12.01 in) probably occurred at Winterbourne Steepleton and Winterbourne Abbas, upstream of Martinstown. (The British record for rainfall in any 24 hours is the 317 mm (12.48 in) recorded at Seathwaite in Cumbria on 19–20 November 2009.)

The Cumbria floods, 2009

Between 18 and 20 November 2009, a deep depression, with a central pressure of 958 millibars at 06.00 GMT on 19 November, was tracking north-east between Scotland and Iceland. A frontal system associated with this depression brought very heavy rain to areas of North Wales, southern Scotland and, in particular, to the Lake District. The heavy rain persisted for about 72 hours, during which it is almost certain that some parts of Borrowdale in Cumbria received no less than 400 mm (15.75 in) of rain. Seathwaite, which is in Borrowdale, set a new British record for the amount of rain received in a 24-hour period with 316.4 mm (12.46 in). (Seathwaite recorded 378 mm (14.88 in) of rain in the 34-hour period from 20.00 GMT on 18 November to 06.00 GMT on 20 November. During that period there were eight consecutive hours when rainfall exceeded 16 mm (0.63 in) per hour.) Exceptional rainfalls were recorded by stations all over the Lake District and it is estimated that the return period for rainfall of such duration is more than 200 years.

The rain led to widespread flooding, especially of the River Derwent, which flows through Borrowdale and reaches the coast at Workington. Here, part of the town was isolated when a road bridge collapsed, leading to the death of a policeman who was attempting to direct traffic away from the dangerous structure. Farther upstream, the worst flooding was in Cockermouth, with more than 2 m (6 ft 6 in) of water in places. The same weather system also caused flooding and disrupted travel in parts of North Wales, in Dumfries and Galloway, and areas in the Borders.

Everything was eternally dreary,
dismal, damned. Even the weather
was insolent and bitchy.

Charles Bukowski, *Ham on Rye*

FASCINATING FACTS:
HOW LARGE ARE RAINDROPS?

Every raindrop is created from millions upon millions of cloud droplets. All cloud droplets are about 20 micrometres (20 μm = 20 thousandths of a millimetre) in diameter, so 1 million are required to make up even a small raindrop, 2 mm (0.08 in) in diameter. Experiments have shown that the largest normal drops are about 6 mm (0.24 in) in diameter. Any larger than this, and their fall speed becomes so great that they tend to break up into smaller droplets. The largest single drops ever recorded (10 mm (0.39 in) diameter) were found in 2004 in Brazil and the Marshall Islands in the Pacific.

It always rains on tents.

Dave Barry

MYTHS AND MISTAKES

An 'upside-down rainbow'

Quite frequently, the media – again! – are full of reports that someone has seen 'an upside-down rainbow'. There's no such thing. Rainbows always occur on the opposite side of the sky to the Sun (or, on rare occasions, the Moon). They are portions of circles, centred on the antisolar point – the point on the sky directly opposite the Sun. If flying in an aircraft, it is sometimes possible to see complete circular rainbows, but still opposite the Sun. In addition – obviously – they only occur when there is rain. Light from the Sun is reflected back by the raindrops and spread out into a spectrum, giving rise to the different colours. Occasionally, a 'secondary' rainbow, with a reversed set of colours may be seen outside the normal 'primary' bow.

But the so-called 'upside-down rainbows' are actually visible high above the Sun – that is, on the same side of the sky. They are an optical phenomenon – of which there are dozens – and are correctly known as 'circumzenithal arcs'. They are caused when sunlight strikes ice crystals floating in the atmosphere. The light passes through the crystals and is spread out into spectral colours. Circumzenithal arcs are one of the brightest and most highly coloured forms of optical phenomenon. They are usually limited to about one third of a circle (120°) that is centred at the zenith, above the observer's head.

Pygmalion *and* My Fair Lady

When Bernard Shaw's play *Pygmalion* was turned into a film, two weather-related phrases were introduced as language exercises for Eliza Doolittle. They were 'The rain in Spain stays mainly in the plain', and 'In Hertford, Hereford and Hampshire, hurricanes hardly ever happen'. These became main songs in the musical *My Fair Lady*. Meteorologically, neither statement is truly correct.

In reality, most of the rain that falls on Spain does not fall in the central plain, but over the mountainous region in the north. And as for hurricanes – they can never happen.

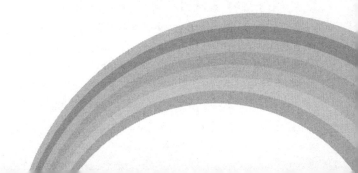

WEATHER LORE

There are a few sayings that contain a grain of meteorological truth, such as:

> **'Rain before seven,**
> **Fine by eleven.'**

Although the times are chosen to rhyme, the idea that rain will last for no more than four hours does have a limited degree of truth. Apart from short-lived showers, most rain comes with frontal systems and, in general, these will pass over within a period of approximately four hours. However, this does not apply to every front. Certain occluded fronts, for example, may sometimes linger over an area, and have even been known to give more-or-less continuous rain for several days.

Another saying is that if there is a ring around the Sun, we will have rain. Similarly:

> **'If the Moon in haloes hides her head,**
> **We shall go wet to bed.'**

When a ring (halo) is seen around the Sun or Moon, it has been produced by ice crystals in a thin sheet of cirrostratus cloud, which usually precedes an approaching warm front. The cirrostratus will thicken into a lower layer of altostratus, which will, in turn, thicken into a still lower sheet of nimbostratus, which will bring rain.

FASCINATING FACTS:
COLD RAIN AND WARM RAIN

You may sometimes hear a meteorologist speak of 'cold rain' or 'warm rain'. This has nothing whatsoever to do with the actual temperature of the rain, but rather describes the way in which it has formed. Raindrops obviously form from cloud droplets, but the latter are so tiny that they float in the atmosphere (actually they do fall, but extremely slowly). But such small droplets rarely collide with one another and combine into larger droplets. Freezing provides the answer. Temperatures high in the atmosphere are so low that ice crystals form (cirrus clouds consist of just ice crystals). Those crystals grow at the expense of surrounding water vapour and cloud droplets until they become heavy enough to start to fall. When they reach warmer layers, the ice melts to give raindrops. This 'ice-crystal' process gives us the 'cold rain'. Most of the rain experienced in Britain is of this type.

This explanation held for many years until it was realised that cloud-top temperatures in many tropical, rain-bearing clouds never dropped below freezing. It was realised that there was a lot of turbulence inside very deep clouds and that would cause cloud droplets to collide and combine. Because the clouds are so deep, droplets continue to grow as they fall, and become raindrops. This 'coalescence' process gives us 'warm rain'. In Britain, warm rain sometimes occurs in summer, when there are very deep clouds, but the tops are not cold enough for freezing to take place.

Bad weather always looks worse through a window.

Tom Lehrer

HIGHS AND LOWS

Wettest place

The location in Britain with the greatest amount of rainfall in a year is Crib Goch, on Snowdon, in Gwynedd, with an average of 4,635 mm (182.42 in) of rain.

Driest place

The driest area of Britain is East Anglia, and the location with the least average annual rainfall is St Osyth in Essex, with just 51.3 mm (2.02 in) of rain.

Dullest place

With an average 736 hours of sunshine per year, Ben Nevis, near Fort William, is the cloudiest place in Britain.

Highest rainfall

The greatest amount of rain to fall in one official, 24-hour rainfall day (beginning 09.00 GMT), was 279 mm (10.98 in), recorded on 18 July 1955, in Martinstown in Dorset.

Highest rainfall in any 24-hour period

On 19 November 2009, Seathwaite, in Borrowdale in Cumbria, recorded 316.4 mm (12.46 in) of rain in a 24-hour period.

Heaviest rainfall in any two-hour period

A total of 155 mm (6.1 in) of rain fell in two hours at Hewenden Reservoir in West Yorkshire on 11 June 1956.

Highest rainfall in one hour

The highest recorded rainfall in a single hour is 92 mm (3.62 in), which fell on Maidenhead in Berkshire on 12 July 1901.

HIGHEST 24-HOUR RAINFALL TOTALS FOR A RAINFALL DAY (09.00–09.00 GMT)

Country	Rainfall (mm)	Date	Location
England	279	18 July 1955	Martinstown (Dorset)
Scotland	238	17 January 1974	Sloy Main Adit (Argyll and Bute)
Wales	211	11 November 1929	Lluest Wen Reservoir (Mid Glamorgan)
Northern Ireland	159	31 October 1968	Tollymore Forest (County Down)

UK RAINFALL RECORDS FOR CONSECUTIVE RAINFALL DAYS (09.00–09.00 GMT)

Days	Rainfall (mm)	Date	Location
Highest two-day total	395.6	18–19 November 2009	Seathwaite (Cumbria)
Highest three-day total	456.4	17–19 November 2009	Seathwaite (Cumbria)
Highest four-day total	495	16–19 November 2009	Seathwaite (Cumbria)

SHORT-DURATION RAINFALL RECORDS

Minutes	Rainfall (mm)	Date	Location
Highest five-minute total	32	10 August 1893	Preston (Lancashire)
Highest 30-minute total	80	26 June 1953	Eskdalemuir (Dumfries & Galloway)
Highest 60-minute total	92	12 July 1901	Maidenhead (Berkshire)

Minutes	Rainfall (mm)	Date	Location
Highest 90-minute total	117	8 August 1967	Dunsop Valley (Lancashire)
Highest 120-minute total	155	11 June 1956	Hewenden Reservoir (West Yorkshire)
Highest 155-minute total	169	14 August 1975	Hampstead (Greater London)
Highest 180-minute total	178	7 October 1960	Horncastle (Lincolnshire)

The rain was steady and unrelenting and, like all steady and unrelenting things, boring.

Ross Thomas, *Briarpatch*

WEATHER WORDS

Báisteach (Irish): Rain.

Blashy (Scots): Rainy.

Bwrw fel o grwc (Old Welsh): Raining as from a bucket.

Bwrw glaw (Welsh): Raining. 'Bwrw' = hitting and 'glaw' = rain, so the literal translation = 'It's hitting rain'.

Dimpsey (Cornish): A dull, wet, drizzly day.

Dish (Scots): Rain heavily.

Dreep (Scots): Steady fall of light rain.

Dribble (Scots): Drizzle.

Driech (Scots): A very common term for anything that is dreary. Frequently used for the weather, it suggests that the weather is grey, damp and depressing, probably overcast and with drizzle. (Also **dree, dreich, dreigh**.)

Droukit (Scots): Drenched (soaked to the skin).

Dubs (Scots): Puddles.

Emptying with rain (Pembrokeshire): Extremely heavy rain.

 Fliuch (Irish): Wet.

Gaeaf gwlyb, mynwent lawn (Old Welsh): Wet winter, a full cemetery.

Gandiegow (Scots): Heavy shower.

Glaw gochel (Old Welsh): Heavy downpour; literally rain to seek shelter from.

Glaw mai, lladd llau (Old Welsh): May rain kill lice.

Glaw Stiniog (Old Welsh): Heavy rain. The slate town of Blaenau Ffestiniog is renowned for its rainfall (a bit like Manchester).

Glawio hen wragedd a ffyn (Old Welsh): Literally this translates as 'raining old wives and sticks' but raining cats and dogs conveys the meaning.

Grumlie (Scots): Unsettled.

Mae'n bwrw glaw (Welsh): It's raining.

Mae'n bwrw hen wragedd a ffyn (Welsh): It's raining very heavily. (Literal meaning: 'It's raining old ladies and sticks'.)

Mizzle (Cornish): An eerie, warm half-mist/half-drizzle.

Muith (Scots): Humid.

Petrichor: A pleasant, distinctive smell that frequently accompanies the first rain after a long period of warm, dry weather in certain regions.

Pipe stapples (Scots): Heavy rain.

Pish-oot (Scots): Downpour.

Plowtery (Scots): Showery.

Plype (Scots): Sudden, heavy shower.

Raff (Scots): Short sharp shower.

Rainin' like a tide (Cornish): A term for very heavy rain.

Scour (Scots): Shower.

Sgrympia Gwyl Grog (Old Welsh): Sudden showers in late September (believed to originate in the Llyn Peninsula).

Shucky (Sussex dialect): A description of unsettled weather.

Smirr (Scots): Light rain.

Smwcian bwrw (Old Welsh): Very fine rain.

Stillicide: A rare, essentially obsolete, English word for a stream of drops, such as raindrops falling from a roof or droplets from the end of an icicle.

Sump (Scots): A great fall of rain.

Swansea rain (Pembrokeshire): Cold rain from the east or south-east.

Thunder-plump (Scots): A sudden, heavy rain shower.

You can't get mad at weather because weather's not about you.

Douglas Coupland

FASCINATING FACTS:
SHOWERS

Although the general public tend to use the term 'shower' for any form of rain, to meteorologists, including TV weather forecasters, it indicates rain from convective clouds (cumulus congestus or cumulonimbus) that is generally short-lived, even though it may be extremely heavy. This is in contrast to the widespread rain that is encountered at frontal systems, which may persist for many hours (and even, in certain cases, for days). After a cold front has passed through there is often a period of individual, isolated showers, each of which may affect only a restricted area.

In general, there are three stages to the development and decay of a shower cloud. In the first stage, which usually lasts about 20 minutes, the cumuliform cloud grows upwards. In the second stage, again about 20 minutes long, heavy rain occurs. This has been produced by coalescence (warm rain) or by the melting of ice produced in the glaciation process (cold rain). Some hail may fail to melt into rain and even persist to reach the ground. It is at this stage that thundery activity may arise. In the final, decaying stage, the rain lessens and any thundery activity decreases. This stage may last for anything from about 30 minutes to perhaps two hours. So the total lifetime of a single shower cell may range from about one hour to two to three hours.

When precipitation starts within a strong shower cloud, the falling particles of rain or hail drag the air downwards creating a cold downdraught within the cloud. Initially the downdraught may co-exist with one or more updraughts within the cloud, but as more precipitation occurs, most of the cloud becomes the site of downdraughts, and the cloud ceases to grow. If the downdraughts are particularly strong, they fan out when they reach the ground, particularly in the direction in which the cloud is moving – ahead of the cloud. This may create a squall line ahead of the cloud, where the cold air, hugging the ground, lifts warmer air above it, which then flows into the cloud, intensifying any updraughts present. Generally the strong downdraughts within the cloud tend to 'quench' the updraughts, leading to the gradual decay of that particular rising cell, whose rain and hail declines. Quite frequently, however, air rising ahead of the main cloud initiates the formation of a new cell, which may itself then develop into a full-blooded shower.

Multicell storms

When a whole cluster of cells are growing at the same time, they may give rise to what is termed a 'multicell storm'. Here, the individual cells may last no more than 45 minutes. When the individual cells are at the stage where they generate lightning, it is often possible to see that there is more than one centre of activity, and each cell will produce activity for about 20 minutes, generally beginning at about the time the initial heavy precipitation reaches the ground. As one cell decays, another may arise and continue the activity, so that the storm as a whole may last for several hours.

Supercells

Supercells are an even more powerful and long-lasting type of storm, and are often associated with extremely violent activity, such as very strong winds, frequent, cloud-to-ground lightning strokes, extensive, damaging hail and even tornadoes. They are characterised by an exceptionally large, rotating updraught, known as a mesocyclone, and have become organised in such a way that they are essentially self-propagating and may last for several hours (commonly six hours or more). The highly organised internal structure of supercells is such that the updraughts and downdraughts are separated, and thus do not interfere greatly with one another. This significantly affects the longevity of the storm. Supercells may arise from multicell storms, in which case they tend to diverge to the right of the overall (gradient) wind. The conditions for their formation not only involve very strong convection, but also strong vertical wind shear – that is, considerable difference in wind speeds with height – and a strong vertical wind veer – a great difference in wind direction with height.

Last night I saw St Elmo's stars,
With their glimmering
lanterns, all at play
On the tops of the masts,
the tips of the spars,
And I knew that we should
have foul weather to-day.

Henry Wadsworth Longfellow,
The Golden Legend

CHAPTER 3

CLOUD, MIST AND FOG

The moral of filmmaking in Britain is that you will be screwed by the weather.

Hugh Grant

WEATHER PEOPLE

Luke Howard (1772–1864): An obsession with clouds

Who would have imagined that a pharmacist and chemist – albeit an industrial chemist – would be so obsessed with observing the weather that, in the early years of the nineteenth century, he came to devise a system of naming clouds that would be accepted worldwide and still be in use today? This was the case with Luke Howard.

Howard (1772–1864) was initially a pharmacist and then became a manufacturing chemist. He was preoccupied with observing the weather – recording weather in the London area from 1801–41 – and particularly in the formation of clouds. His essay, *On the Modifications of Clouds*, was presented to the Askesian Society in 1802 and published in 1803. In it he proposed a formal classification of clouds, similar to that used previously by Carl Linnaeus for plants (in 1753) and animals (in 1758). Although not the first to suggest such a cloud classification scheme, his proposals were widely accepted and some of his terms (such as cirrus, cumulus, nimbus and stratus) remain in use today. The classification scheme was successful because of his use of Latin terms and also his recognition of (and allowance for) changes in the form and structure of clouds. Previously, clouds had been generally regarded as so changeable that they would defy any proper classification. His scheme was rapidly and widely adopted, not only by those interested in natural philosophy – as scientific subjects were then generally known – but also by writers and poets (such as Goethe and Shelley), and artists (in particular, by Constable, who made numerous studies of clouds, annotating them with details of the prevailing weather).

Although Howard's initial scheme has been somewhat modified in more recent years, the basic concepts remain true. For this reason, and the fact that he studied the weather for many decades, Howard is often known as 'the Father of Meteorology'. According to English Heritage, the blue plaque marking the house where he lived has the most admired inscription of any blue plaque in London – 'Namer of clouds'.

I am the daughter of Earth and Water
And the nursling of the Sky;
I pass through the pores of the oceans
and shores.
I change, but I cannot die.
For after the rain, when with never a stain
The pavilion of Heaven is bare,
And the winds and sunbeams with their
convex gleams
Build up the blue dome of air,
I silently laugh at my own cenotaph,
And out of the caverns of rain,
Like a child from the womb, like a ghost
from the tomb,
I arise and unbuild it again.

Percy Bysshe Shelley, 'The Cloud'

Charles Thomson Rees Wilson (1869–1959): Life-changing experiences on Ben Nevis

A short spell at the Ben Nevis Observatory in 1894, as a temporary relief for one of the permanent staff, prompted Charles Wilson to investigate phenomena that he had observed during his short stay. He was particularly struck by the appearance of coronae (coloured rings seen around the Sun) and a glory (a set of coloured rings that appear around the shadow of an observer's head). Both of these occur in water-droplet cloud. Another incident saw him caught in a thunderstorm on Ben Nevis and experiencing the electrical forces involved.

Wilson became a distinguished physicist. Born at Glencorse in Midlothian, he took his BSc at Manchester University, then gained a scholarship to Sidney Sussex College, Cambridge, and was awarded first-class honours in the natural science tripos. After a short period as a teacher, he returned to Cambridge in 1894 to work at the Cavendish Laboratory. Prompted by his observations on Ben Nevis, he carried out a series of experiments to establish the condensation mechanisms in clouds, devising apparatus that allowed the formation of condensation droplets under controlled conditions. He realised that, under specific conditions, condensation could be initiated by the passage of charged particles through the chamber. It was at the Cavendish that J. J. Thomson carried out the fundamental experiments that determined the existence of the electron in 1897. He applied Wilson's cloud-chamber technique to determine the charge on the electron.

Thomson was awarded the Nobel Prize for Physics in 1906 for the discovery of the electron.

Wilson turned to the study of ions present in the air and suspected the existence of what are now known as cosmic rays. His own results were inconclusive, and cosmic rays were not discovered until 1913 by Victor Francis Hess. From 1903, Wilson studied atmospheric electricity, but returned to his cloud-chamber work in 1910. His refinements eventually led to his being able to detect the track of a single ionising particle through the chamber. This work proved to be a vital advance in the study of atomic physics. He was awarded the Nobel Prize in Physics in 1927.

> The weather and love are
> the two elements about which
> one can never be sure.
>
> Alice Hoffman, *Here on Earth*

CLOUDS AS AN ART FORM

Around the beginning of the nineteenth century, artists began to paint more realistic skies – where meteorologists can even recognise specific cloud types – rather than the highly stylised and largely symbolic clouds that had appeared previously. One who took a great interest in the sky and painted realistic clouds was **John Constable** (1776–1837). He probably developed a particular feeling for wind and sky when, for a short period, he worked at the family windmill, before devoting his life to painting.

Constable may be best known for his paintings of Suffolk ('Constable Country') with works such as *Dedham Vale* and *The Hay Wain*, but he was fascinated by clouds and the weather, and with painting them – he termed it his 'skying'. He considered the sky to be the key to successful landscape painting. He made dozens of sketches of clouds – most showing just clouds, with no landscape – annotating many with precise details of the prevailing weather and the changes that occurred later in the day. He was fully aware of Howard's cloud-classification scheme and often gave details of the wind speed using the Beaufort Scale. Although he started to give details of the weather in 1806 with sketches made in the Lake District, he intensely studied and sketched the sky during 1820–22, when staying on Hampstead Heath, where he was in an ideal position to study the sky. (No less than 54 of these studies still exist.) His notes are so accurate that they have been used by modern-day meteorologists to determine, not only the weather on certain particular days, but also the development of weather systems over a period. His *Hampstead Heath with a Rainbow* and its naturalistic depiction of cumulus and cumulonimbus clouds was painted later, in 1836.

Joseph Mallord William Turner (1775–1851) is known for the dramatic skies in his paintings, rather than for accurate depiction of clouds, unlike other painters active around the same time, such as Richard Bonington and John Constable. Turner's skies are often highly dramatic and impressionistic, rather than naturalistic, as shown, for example, by his *Snow Storm: Steam-Boat off a Harbour's Mouth*, of 1842. His sunrise, sunset and other skies are frequently highly coloured (such as his famous *Rain, Steam and Speed – The Great Western Railway* of 1844) and it has been suggested that these were influenced by the strongly coloured skies that he saw in the early part of the nineteenth century and which were produced by the series of large volcanic eruptions (particularly the immense eruption of Tambora, Indonesia, in 1815 which ejected large amounts of material high into the atmosphere and produced dramatic sunrise and sunset scenes around the world).

Fog everywhere. Fog up the river, where it flows among green aits and meadows; fog down the river, where it rolls defiled among the tiers of shipping and the waterside pollutions of a great (and dirty) city. Fog on the Essex marshes, fog on the Kentish heights. Fog creeping into the cabooses of collier-brigs; fog lying out on the yards and hovering in the rigging of great ships; fog drooping on the gunwales of barges and small boats. Fog in the eyes and throats of ancient Greenwich pensioners, wheezing by the firesides of their wards; fog in the stem and bowl of the afternoon pipe of the wrathful skipper, down in his close cabin; fog cruelly pinching the toes and fingers of his shivering little 'prentice boy on deck. Chance people on the bridges peeping over the parapets into a nether sky of fog, with fog all round them, as if they were up in a balloon and hanging in the misty clouds.

Charles Dickens, *Bleak House*

NOTABLE WEATHER EVENTS

The Great Smog, 1952

In early December 1952 a slow-moving anticyclone lay over the British Isles, bringing cold weather and windless conditions. London lay beneath a temperature inversion (temperature became warmer with height, rather than colder), suppressing the normal turnover of air. Smoke from the innumerable coal fires, which were being used to keep warm, together with numerous other forms of pollution, such as vehicle exhaust and industrial emissions, was trapped over the city. With no wind to disperse the fumes, a dense layer of smog was created over London. It became particularly dense on 5 December and lasted until 9 December. The smog was so dense that it penetrated into buildings, and various theatres and cinemas had to be closed, because the audiences were unable to see the stage or screen. Transport was badly affected and only the Underground continued to function. Even walking became almost impossible and street lights failed to penetrate the murk.

At the time the event was largely tolerated, because Londoners were accustomed to dense fogs, but subsequent analysis suggested that some 4,000 people had died prematurely and some additional 100,000 suffered respiratory illness because of the smog. (A much later analysis suggests that the number of excess deaths was even greater – perhaps as many as 12,000.) These findings were primarily responsible for the Clean Air Act, which was eventually passed in 1956, with the introduction of controls on emissions, 'smokeless' fuels, and the increased use of gas and electricity for central heating – practically unknown in 1952. The incidence of such events has decreased, although there was a similar smog ten years later, in 1962. Since then, however, the infamous London 'pea-soupers' have become a thing of the past. Although dense fogs do still occur, they are 'white' fogs (consisting of largely water vapour alone) and do not contain such high levels of injurious pollutants.

MYTHS AND MISTAKES

'The Island Cloaked in Mist'

One theory for the origin of the name of the Isle of Man (*Mann* in Manx Gaelic) is that it comes from the name of Manannán mac Lir, a mythological Celtic sea god. When the island was threatened by invaders, Manannán mac Lir would climb to the top of Snaefell, the highest mountain on the island (620 m/2,034 ft), and cast his misty cloak around the island, hiding it from view.

The Brocken Spectre

The Brocken Spectre (also sometimes known as the 'Brocken bow' or 'mountain spectre') is an apparently enormous shadow of the observer, cast on to mist or the top of a layer of cloud that lies opposite the Sun. It is commonly observed from mountains or suitable sites – it has been observed from the Golden Gate Bridge in California and from aircraft – but is named after the

Brocken, a peak in the Harz Mountains in Germany. But the apparent magnification of the image is an optical illusion that is related to one called the 'Tunnel Illusion'. The converging lines of shadow create the impression that the shadow is closer and larger. In addition, the fact that the shadow is created at different distances in the bank of mist or cloud, makes it impossible for the observer to focus on any sharp details, increasing the appearance that the shadow is greatly enlarged.

CLOUD TYPES

FASCINATING FACTS:
TYPES OF CLOUD

There are ten main types of cloud. Each genus is normally found at a particular range of heights, which, by convention, are described as 'high', 'middle' and 'low'. However, one genus, cumulonimbus, may have a very low base but stretch through all three levels. The names of the types are given here.

Altocumulus

A layer of cloud at middle levels, consisting of individual rolls or plates of cloud, which show distinct darker shading and have clear sky between them. A beautiful form of cloud, often described as forming 'a mackerel sky'. The individual patches of cloud are between 5° and

1° across, measured 30° above the horizon, and thus between the sizes of stratocumulus and cirrocumulus.

Altostratus

A relatively featureless layer of middle-level, grey cloud, through which the Sun, if visible, appears as if through ground glass and casts no shadows. It often thickens and becomes nimbostratus.

Cirrocumulus

A high-level cloud, consisting of a layer of tiny, pale, white or bluish plates of cloud, with no shading. When particularly thin, the cloud may be difficult to see. The individual cloudlets are small, being always less than 1° across, measured 30° above the horizon, and are thus smaller than altocumulus.

Cirrostratus

A featureless, thin sheet of ice-crystal cloud at a high level. It is often difficult to see, but frequently displays a whole range of optical phenomena (which may be coloured arcs or white spots of light and circular haloes).

Cumulus

A rounded heap cloud, found at low levels and usually building up during the day. It may grow upwards and eventually become a cumulonimbus cloud.

Cirrus

A high-level, ice-crystal cloud, usually consisting of thin wisps or trails of cloud, one form of which is commonly called 'mares' tails'. Thicker patches may be dense enough to hide the Sun and often form the tops of cumulonimbus clouds.

Cumulonimbus

A large, towering cloud that extends to very great heights, often with a flattened top, or 'anvil' of cirrus. The base is often very dark and ragged and produces heavy showers of rain. It may also give rise to lightning and hail.

Nimbostratus

A dark grey cloud, generally at middle levels, but which may extend down towards the surface. It gives rise to prolonged periods of rain.

Stratocumulus

A low-level layer of individual plates or 'pancakes' of cloud, with dark shading and breaks of clear sky between them. The individual cloudlets are more than 5° across, measured 30° above the horizon. Only occasionally will stratocumulus produce a little light rain.

Stratus

A low-level layer of relatively featureless, grey cloud, which may shroud the tops of high buildings or hills or even reach the ground (when it appears as fog). It may rarely give rise to drizzle or produce a fall of minute ice crystals.

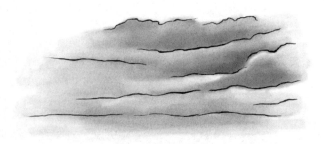

WEATHER LORE

There are a number of pieces of weather lore that relate to clouds. Some of these contain a grain of truth, but most need to be qualified, because they may not apply in every circumstance.

> 'Mares' tails and mackerel scales
> make tall ships carry low sails.'

This is often true. The 'mares' tails' are high cirrus clouds that have been shaped by the upper winds. Cirrus clouds may signal an approaching frontal system, with its increasing winds. However, wisps of cirrus cloud are often seen at other times, so it is not an infallible warning.

The term 'mackerel scales' or 'mackerel sky' is often applied to cirrocumulus or altocumulus clouds that consist of individual clumps of cloud. Such clouds are sometimes a sign of an advancing low pressure system, but at others may appear behind a retreating system, so their use as a warning sign needs to be accompanied by other signs of a deterioration in the weather.

> 'If woolly fleeces bestow the heavenly
> way, be sure no rain will come today.'

Scattered cumulus clouds that appear like fluffy sheep are often a sign of settled weather. Indeed they are known as 'fair-weather clouds', or 'fair-weather cumulus'. This piece of lore is true, provided the clouds remain small. But later in the day, the clouds may grow into high cumulus congestus or even cumulonimbus, both of which may result in rain.

> **'When clouds look like black smoke
> a wise man will put on a cloak.'**

Large clouds such as cumulus congestus or cumulonimbus are laden with large quantities of large water droplets. Their bases, in particular, look much darker – sometimes almost black – than smaller tufts of cumulus cloud, which generally have much smaller water droplets and are not as dense, so the white light scattered by the cloud droplets can escape, making them look much lighter in colour. Very dark clouds are an almost certain sign of imminent rain.

> **'The more cloud types present, the
> greater the chance of rain or snow.'**

Different cloud types may indicate instability in higher regions of the atmosphere, bringing with it the right conditions for storms. This is particularly true when there are high jet-stream clouds with increasing cirrus at a lower level, beneath which are smaller cumulus clouds. In many cases, the motion of the clouds will be in different directions, reflecting the differing winds at various heights.

> **'A summer fog for fair, a winter fog for rain. A fact most everywhere, in valley or on plain.'**

Fog is formed, when the temperature falls to the dew point, at which humidity rises to 100 per cent. In summer, this will only happen on clear nights. Clouds act as a blanket at night, preventing heat from the ground from being radiated away to space. The air temperature does not cool to the dew point. Fog is also formed when warm, moisture-laden air blows over a cold surface, such as the sea. This produces sea fog, which may be carried on to nearby coasts by the wind.

> **'The higher the cloud the better the weather.'**

This saying is false. Cirrus clouds are typically found at altitudes of around 9 km (30,000 ft). When they are associated with jet streams they often indicate the approach of bad weather.

FASCINATING FACTS:
'NIGHT-SHINING' CLOUDS

On rare occasions, for about six weeks on either side of the summer solstice (21 June), and for observers north of about latitude 45°N, bright, 'electric-blue' clouds may be seen shining in the direction of the North Pole around midnight. These clouds are known as noctilucent clouds ('night-shining clouds') and are the highest clouds ever to occur in the atmosphere, far above other clouds, even nacreous clouds. Typically their altitudes are 80–85 km (50–53 miles), and their existence poses considerable problems. They look somewhat like the much lower cirrus clouds and, like them, are known to consist of ice, but there is still disagreement over whether the water comes from the Earth's surface, or from outside the atmosphere, introduced by cometary material. The generally favoured theory at present is that the water is derived from the gas methane, which is released from the surface and may be transported to such great heights. Similarly, because tiny solid particles (nuclei) are required, on which ice crystals may form, these nuclei may be micrometeorites from space, clusters of ions formed by cosmic rays (also from space, of course), or minute particles of volcanic or other dust, lofted by some unknown mechanism to great altitudes.

There are suggestions that these noctilucent clouds are becoming more frequent, and spreading to lower latitudes, but it is quite uncertain whether they are actually occurring more frequently, or whether more people are taking note of them. They were not reported before 1884, after the gigantic volcanic eruption of Krakatoa in 1883, when people paid particular attention to the sky because of the spectacular sunsets, but it appears difficult (if not impossible) for any volcanic eruption, however powerful, to send ash to such extreme heights.

> Dost thou know the
> balancings of the clouds,
> the wondrous works of him which
> is perfect in knowledge?
>
> Job 37:16

HIGHS AND LOWS

Cloudiest month

Greater London didn't have a single hour of sunshine during December 1890.

Foggiest place

Between 1963 and 1976, Great Dun Fell in Cumbria at an altitude of 857 m (2,812 ft), experienced an average of 233 days of fog at 09.00 GMT, which may be compared with the average of six days for the period 1961–1990 at Carlisle in Cumbria, at an altitude of 26 m (85 ft).

Places with least fog

Two locations (in particular) in the British Isles experience a very low incidence of foggy days. They are Calshot, at the mouth of Southampton Water in Hampshire, and Lossiemouth, in Moray in Scotland. It was for this reason that the two sites were specifically chosen as the bases for flying boats. Calshot, in particular, was an important flying-boat base in World War Two. Shannon in Ireland also had a low incidence of fog, and was used as the base for the first flying-boat passenger service across the Atlantic between Europe and North America, established after the war.

Lift up your lovely heads and look
As wind turns clouds into a picture book.

Mary O'Neill, *Wind Pictures*

WEATHER WORDS

Barber (Scots): Freezing mist.

Broch (Scots): Although broch is normally a term for an Iron Age, circular, stone building that served as a fortified home and which is found in the northern parts of Scotland, it is also used for a ring around the Moon – a lunar halo. This (together with a similar solar halo around the Sun) occurs when a thin sheet of cirrostratus (ice-crystal cloud) covers the sky.

Call-boys (Sussex): A term for tiny patches of cloud that drift upslope, usually heralding the morning break-up of fog over low-lying ground.

Clag (Pembrokeshire): Fog. Also 'Clagged in' = foggy.

Cymylau blew geifr (Old Welsh): Cirrus, streaky clouds. Literally translates as goats' hair clouds.

Cymylau mynydde (Old Welsh): Clouds that appear in the distance like mountains. Literally, mountain clouds.

Cymylau boliog (Old Welsh): Literally 'fat clouds', cirrostratus clouds.

Cymylau caws a llaeth (Old Welsh): Literally 'cheese and milk clouds', mackerel sky, cirrocumulus clouds.

Cymylau caws a maidd (Old Welsh): Literally 'cheese and whey clouds', again cirrocumulus clouds.

Cymylau pysgod awyr (Old Welsh): Literally 'clouds like sky fish', essentially long stripes of cloud that appear like large fish.

Drackie (Scots): Misty.

Drumkie (Scots): Cloudy.

Haar: A widespread term for a damp sea fog that invades the land from the North Sea, often carried inland by a sea breeze. Originally used from Lincolnshire northwards to Scotland, it is now often used anywhere on the east coast, including in East Anglia.

Port-boys (Sussex dialect): Small, low clouds in a clear sky.

Rouline (Scots): Damp, misty.

Smoch (Scots): Thick fog.

Ure (Scots): Damp mist.

Windogs (Sussex dialect): White clouds blown by the wind.

Is not their climate foggy,
raw, and dull?
On whom, as in despite,
the sun looks pale,
Killing their fruit with frowns.

William Shakespeare, *Henry V*

FASCINATING FACTS:
MOTHER-OF-PEARL CLOUDS

Occasionally, one type of rare cloud gives such dramatic displays that they are widely reported on the television news and in other media. These are the mother-of-pearl clouds also known as nacreous clouds. (The formal meteorological name is 'polar stratospheric clouds'.) As their common name suggests they display a wonderful range of colours and, because they occur just after sunset (or before sunrise) appear against a dark sky, which makes them all the more striking. They are ice-crystal clouds that occur in the lower portion of the stratosphere – where clouds are rare – at altitudes of 15–30 km (50,000–100,000 ft). The beautiful colours arise from the phenomenon of diffraction, where light from the Sun (which is below the horizon for observers on the ground but still illuminates the clouds) is bent around the particles. A particular shade is an indication that all the particles causing that colour are of the same size. They are also a form of wave cloud, caused by air flowing over mountains or other obstacles on the ground that produce waves (even at high altitudes).

There is no way that we can predict the weather six months ahead beyond giving the seasonal average.

Stephen Hawking,
Black Holes and Baby Universes

CHAPTER 4
WIND

I cannot command winds and weather.

Horatio Nelson

WEATHER PEOPLE

Edmond Halley (1656–1742): Not just an astronomer

Shortly after the formation of the Royal Society in 1660 as the 'Royal Society of London for Improving Natural Knowledge', with a charter from King Charles II, several of its Fellows were active in numerous different fields of study. Some who were involved in investigations of the weather included Christopher Wren, Robert Hooke, Isaac Newton and Edmond Halley, who should be known for his many other contributions as well as for 'his' comet.

Halley was an English astronomer and mathematician who made extremely significant contributions to many different fields of the natural sciences. He is perhaps best remembered for his astronomical work, particularly the determination of the orbit of the comet named after him, and for predicting the date of its return (although this occurred after Halley's death). He also prompted (and paid for) the publication of Isaac Newton's fundamentally important *Philosophiæ Naturalis Principia Mathematica* (1687) known simply, to this day, as the *Principia*. He briefly commanded the pink HMS *Paramour* – a small three-masted vessel, the first built specifically for research – on three voyages in the Atlantic. One specific meteorological contribution was his chart of winds (including the trade winds) published in 1686, which was followed in 1701 by his chart of magnetic deviation. This chart was the first ever to make use of isolines (lines joining points at which a particular physical property has the same numerical value), now most commonly seen in meteorology as isobars (joining points with the same pressure).

The Halley Trade Wind Chart

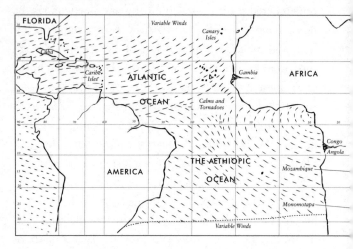

Rear Admiral Sir Francis Beaufort (1774–1857): Taming the winds

Anyone who has ever listened to the Shipping Forecast will have heard the strength of the wind referred to by some such terms as 'north-west Force 4 or 5, backing westerly'. Wind force, familiar to all sailors, refers to the numerical scheme devised by Francis Beaufort that came to be generally accepted.

Beaufort was a British naval officer who, in 1806 when still a commander, devised a means of measuring (or more strictly speaking, estimating) the strength of the wind at sea, initially based upon the sails that a frigate could carry under different conditions. It was not until 1838 that Beaufort's scheme was officially

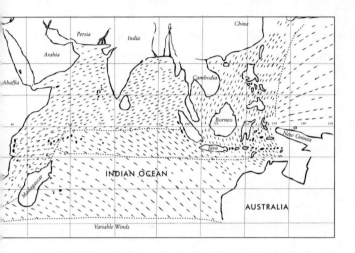

adopted by the Royal Navy. Although named after Beaufort, the concept of a numerical scale had been employed previously by Alexander Dalrymple, Hydrographer to the East India Company, and by the famous civil engineer, James Smeaton. Some of the terms had even been in use since the seventeenth century, and were included in Daniel Defoe's book *The Storm*, which described the violent storm of 1703. Beaufort's innovation was to use a scale to classify the force (however great or small) exerted by the wind. With modifications, including its expansion to conditions on land, the scale remains in use today. He also introduced a series of letters to indicate past and present weather, and the use of symbols to give a graphical method of plotting observations.

In 1829 Beaufort was appointed head of the Admiralty's Hydrographic Office, a post that he held for 25 years. He was responsible for it becoming the world's leading hydrographic organisation, with Admiralty charts renowned for their accuracy and for their use worldwide. Beaufort was involved in, and made major contributions to, many scientific fields including geography, geodesy, oceanography and astronomy, as well as meteorology.

FASCINATING FACTS: THE BEAUFORT SCALE

The Beaufort Scale is a standard scale, running from Force 0 to Force 12, used to give an indication of the wind speed. The original scale applied to conditions at sea, but was subsequently modified for use on land. All of the descriptions were generalised to be universally applicable.

THE BEAUFORT SCALE (FOR USE AT SEA)

Force	Description	Sea state	Speed	
			Knots	m/s^{-1}
0	Calm	Like a mirror	<1	0.0–0.2
1	Light air	Ripples, no foam	1–3	0.3–1.5
2	Light breeze	Small wavelets, smooth crests	4–6	1.6–3.3
3	Gentle breeze	Large wavelets, some crests break, a few white horses	7–10	3.4–5.4
4	Moderate breeze	Small waves, frequent white horses	11–16	5.5–7.9
5	Fresh breeze	Moderate, fairly long waves, many white horses, some spray	17–21	8.0–10.7
6	Strong breeze	Some large waves, extensive white foaming crests, some spray	22–27	10.8–13.8
7	Near gale	Sea heaping up, streaks of foam blowing in the wind	28–33	13.9–17.1

THE BEAUFORT SCALE (FOR USE AT SEA) CONTINUED

8	Gale	Fairly long and high waves, crests breaking into spindrift, foam in prominent streaks	34–40	17.2–20.7
9	Strong gale	High waves, dense foam in wind, wave crests topple and roll over, spray interferes with visibility	41–47	20.8–24.4
10	Storm	Very high waves with overhanging crests, dense blowing foam, sea appears white, heavy tumbling sea, poor visibility	48–55	24.5–28.4
11	Violent storm	Exceptionally high waves may hide small ships, sea covered in long, white patches of foam, waves blown into froth, poor visibility	56–63	28.5–32.6
12	Hurricane	Air filled with foam and spray, visibility extremely bad	>64	>32.7

THE BEAUFORT SCALE (ADAPTED FOR USE ON LAND)

Force	Description	Events on land	Speed	
			Km/h⁻¹	m/s⁻¹
0	Calm	Smoke rises vertically	<1	0.0–0.2
1	Light air	Direction of wind shown by smoke but not by wind vane	1–5	0.3–1.5
2	Light breeze	Wind felt on face, leaves rustle, wind vane turns to wind	6–11	1.6–3.3
3	Gentle breeze	Leaves and small twigs in motion, wind spreads small flags	12–19	3.4–5.4
4	Moderate breeze	Wind raises dust and loose paper, small branches move	20–29	5.5–7.9
5	Fresh breeze	Small leafy trees start to sway, wavelets with crests on inland waters	30–39	8.0–10.7
6	Strong breeze	Large branches in motion, whistling in telephone wires, difficult to use umbrellas	40–50	10.8–13.8

THE BEAUFORT SCALE (ADAPTED FOR USE ON LAND) CONTINUED

7	Near gale	Whole trees in motion, difficult to walk against wind	51–61	13.9–17.1
8	Gale	Twigs break from trees, difficult to walk	62–74	17.2–20.7
9	Strong gale	Slight structural damage to buildings; chimney pots, tiles, and aerials removed	75–87	20.8–24.4
10	Storm	Trees uprooted, considerable damage to buildings	88–10	24.5–28.4
11	Violent storm	Widespread damage to all types of building	102–117	28.5–32.6
12	Hurricane	Widespread destruction, only specially constructed buildings survive	>118	>32.7

Who has seen the wind?
Neither I nor you.
But when the leaves hang trembling,
The wind is passing through.
Who has seen the wind?
Neither you nor I.
But when the trees bow down their heads,
The wind is passing by.

Christina Rossetti, 'Who Has Seen the Wind?'

NOTABLE WEATHER EVENTS

The Tay Bridge Disaster, 1879

Although the primary causes of the collapse of the Tay Bridge on 28 December 1879 – which saw the train crossing at the time plunge into the depths below, killing everyone on board – were undoubtedly the weakness of many bridge components and a failure to take sufficient wind loading into account, it is almost certain that other aspects of the weather were also closely involved. It is now believed that stormy weather earlier in the year had partially weakened the structure; wind speeds of about 62 knots (114 kph) were recorded that day at Glasgow and Aberdeen. Recent estimates of the wind speed at the time of the bridge's failure suggest that it was gusting to about 70 knots (130 kph). In addition, there are reports of two or three waterspouts having been seen near the bridge at the time of the disaster, although it is not known if they struck the bridge. Although early reports suggested a death toll of 75, and this figure is widely quoted, modern research suggests that the number of deaths was actually 59 and only 46 bodies were ever recovered.

The disaster was, of course, the subject of no less than three famous poems by William McGonagall, widely regarded as the worst poet ever to write in English. He incorrectly quoted the number of deaths as 90, and went on to write another (dreadful) poem about the replacement bridge that was opened in July 1887.

The East Coast Floods, 1953

On Saturday, 31 January 1953, a fairly deep depression moved down the North Sea from Scotland towards the Netherlands. It was accompanied by very strong winds (gale to severe gale force) on the western side of the low's centre. These winds affected all the eastern counties of Scotland and England and created a storm surge of water that travelled down the North Sea. The combination of the low pressure, the wind and high tide raised water levels by as much as 5.6 m (18 ft) in places, overtopping almost all flood defences during the night of 31 January to 1 February. There were devastating floods in East Anglia and the Thames Estuary, with a total of 307 fatalities in Lincolnshire, Norfolk, Suffolk and Essex, and 19 in Scotland. The situation in the Netherlands was even worse, with 1,836 deaths, and there were 28 fatalities in Belgium. Although London escaped flooding, a report on the devastation caused by the storm surge acted as a stimulus to the creation of the Thames Barrier, and also to the erection of a tidal barrier at the mouth of the River Hull, where it enters the Humber Estuary. In the Netherlands it led to the development of the vast Delta Plan of dams and tidal barriers. A somewhat similar surge on 16–17 February 1962 affected northern Germany, with severe flooding in Hamburg, where 315 people lost their lives.

The loss of the Princess Victoria, 1953

The North Sea was not the only area affected by the deep depression of 31 January 1953 and its accompanying gale-force winds. Earlier that day, conditions to the west and north of Scotland were extremely severe and many vessels experienced difficulties. On the morning of 31 January at 07.45 GMT, the MV *Princess Victoria* left Stranraer (in what was then Wigtownshire, but is now Dumfries and Galloway) for Larne in Northern Ireland. On leaving the sheltered waters of Loch Ryan, the ferry encountered heavy seas which damaged the stern doors, allowing water to flood the car deck. An attempt to return to Loch Ryan was unsuccessful, so the decision was taken to try to reach Northern Ireland. The ferry started to list, because of shifting cargo, and continued to take on water. A distress call was sent out at 09.46 GMT and an SOS at 10.32 GMT. The order to abandon ship was given at 14.00 GMT. The final radio message at 13.58 GMT reported the ship on its beam ends and it is assumed to have sunk shortly afterwards. Royal Navy vessels attempted to provide assistance, and several merchant vessels put to sea from Northern Ireland, as did the Portpatrick and Donaghadee lifeboats, but the ferry's position was uncertain until it reported it had sighted Northern Ireland. (Unfortunately, the only aircraft available had been assisting at rescues off the isles of Lewis and Barra and did not arrive on the scene until after the ferry had sunk.) The large vessels were unable to rescue survivors from the lifeboats because of the exceptional seas that threatened to smash them into the larger ships. Only when the Donaghadee lifeboat, the *Sir Samuel Kelly*, arrived could anyone be rescued. There were 44 survivors, but 133 lives were lost, including all the ship's officers. It was the worst British maritime disaster since World War Two. Numerous other vessels, particularly many fishing trawlers and their crews, were lost in the storm.

The Irish Sea

EDINBURGH
GLASGOW
NORTH CHANNEL
STRANRAER
LARNE
BELFAST
ISLE OF MAN
WARREN POINT
IRISH SEA
LIVERPOOL
DUBLIN
ST GEORGE'S CHANNEL
FISHGUARD
SWANSEA
BRISTOL

The collapse of the Ferrybridge cooling towers, 1965

On 1 November 1965, exceptionally high winds were experienced across most of Britain, with a hurricane-force gust of 102 knots (188 kph) recorded in Lanarkshire. At sea, five Norwegian sailors were lost from a Norwegian frigate off Malin Head in County Donegal. Conditions were particularly severe across Central Scotland and in northern England, where there were numerous injuries and a number of fatalities. The wind was extreme even in southern England, with injuries in London and Essex.

The most spectacular effect, however, was the collapse of three giant cooling towers at Ferrybridge Power Station, near Doncaster in South Yorkshire. The whole set of eight towers had been completed in 1961, and, at 114.3 m (375 ft) high, were the tallest in Europe. Each tower weighed 8,000 tons. Unfortunately, errors had been made in the calculations and wind-tunnel tests. The strength of the upper sections of the towers was inadequate and tests had been carried out on a single model tower, so the funnelling effect and turbulence caused by the second row of towers had not been taken into account. Three of the four towers on the downwind side collapsed into a shower of dust and debris; the fourth tower had a very large crack, but remained standing. The other four towers to windward were also damaged. Fears were expressed for the two giant chimneys 198 m (650 ft) high, but they were undamaged. Luckily, it was 'tea break' and no workers were inside the towers, so there were no casualties or severe injuries.

FASCINATING FACTS:
THE JET STREAM

Television and radio weather forecasters often mention the jet stream when reporting on the weather over the British Isles. There are actually several jet streams, but the one that has the dominant effect on British weather is the Polar Front Jet Stream. This is a generally westerly, high-speed airflow at an altitude of approximately 10 km (30,000 ft). The lower limit in speed for jet streams is taken as 61 knots (110 kph). Although generally westerly, the jet steam has waves in latitude (known as Rossby waves) that mean the flow is frequently south-westerly or north-westerly. On occasions, the flow may be roughly meridional (that is, northerly or southerly) and, rarely, even easterly. When the jet stream is strong, the meanders are reduced in size, but when the speed is low, the excursions north and south may be very large indeed. The waves tend to steer depressions across the country and draw warm air north and cold air south. The resulting weather will depend on the exact position and motion of the waves (which is generally eastwards). At times, the meanders of the jet stream cease to progress eastwards and may become stationary, in what is known as a 'blocking' situation, bringing essentially unchanging weather conditions to the country below, sometimes for days or even weeks on end.

Scotland! thy weather's like a modish wife;
Thy winds and rain for ever are at strife.

Aaron Hill

MYTHS AND MISTAKES

It is not the Gulf Stream

It seems to be widely believed that the mild weather over the British Isles, when compared with conditions at the same latitude on the other side of the Atlantic, is thanks to the Gulf Stream. But that's not quite right. The Gulf Stream is the current of warm water that emerges from the Gulf of Mexico and passes north along the east coast of the United States. But the actual Gulf Stream only reaches as far as Cape Hatteras. There the flow of water turns east and becomes known as the North Atlantic Current. Part-way across the Atlantic, that current sends off a branch (the North Atlantic Drift) towards the north-east, and that eventually penetrates into the Arctic Ocean north of Norway. The prevailing winds (from the west), passing over the North Atlantic Drift, pick up warmth from the warm water, and it is that air that produces the mild climate of the British Isles (and much of the rest of western Europe).

The Gulf Stream

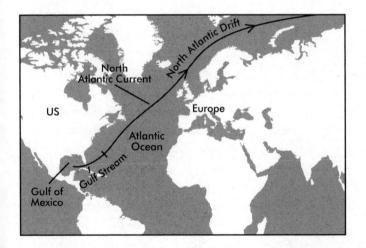

FASCINATING FACTS:
WHAT IS A 'STING JET'?

A term that you may occasionally hear weather forecasters use, when predicting or discussing violent winds, is that they are caused by a 'Sting Jet'. This describes a phenomenon, present in some deep depressions, that has only been recognised in recent decades. Extremely high winds sometimes occur just behind the cold front of the developing depression. The most powerful winds occur in a very small region, just a few tens of kilometres wide, close to the end of the curl of cloud that wraps partially round the low-pressure centre. What happens is that cold dry air from a relatively high altitude (3–4 km (approximately 10,000–13,000 ft)) descends towards the surface. As this happens, rain or snow falling into that descending air evaporates, causing the air to cool and thus become even denser. The air accelerates and reaches the surface as extremely damaging winds. These only persist for a relatively short period of time, until the band of cloud (and the jet of cold air within it) wraps completely round the low-pressure centre. It was just such a phenomenon that produced the most damaging winds during the great storm of 16 October 1987, which actually led to the discovery of the effect.

Grumphie smells the weather,
An' grumphie sees the wun'.
He kens when clouds will gather
An' smoor the blinkin' sun.
Wi' his mou' fu' o' strae
He to his den will gae.
Grumphie is a prophet
Wet weather will we hae.

Galloway rhyme about pigs, which are thought to be able to see the wind

WEATHER LORE

'A high wind drives away the frost.'

Frost tends to form when the air is still and there is no cloud cover at night. The ground radiates heat away to space, causing the temperature to drop and water vapour to be deposited as ice crystals. However, even if there is a moderate wind, the air is mixed throughout the lowermost layer of the atmosphere and temperatures rarely drop low enough for water to freeze. Similarly, if a wind arises after frost has formed, it may mix warmer air from a higher layer of the atmosphere with the lowermost air, warming the surface and causing the frost crystals to disappear.

FASCINATING FACTS:
THE HELM WIND

Rather surprisingly, there is just a single named wind in the British Isles. This is the Helm wind, a particularly violent wind that sometimes affects the area of the Eden Valley in Cumbria. It is a very strong, cold wind that cascades down the south-western escarpment of Cross Fell and occurs most frequently in late winter and spring when the gradient wind is from the north-east. (Cross Fell is the highest region of the Pennines and consists of an upland plateau, prone to being blanketed in hill fog, where snow is often extremely persistent.) Apart from being cold, the wind is exceptionally gusty and produces a loud shrieking noise. It may have a strong desiccating effect and may persist for days. It is often accompanied by a cap or dense line of clouds above Cross Fell, known as the 'Helm', which may predict the onset of the Helm wind itself. In addition, there is sometimes a long, almost stationary roll of cloud, the 'Helm Bar', which forms downwind, 1–6 km (3,280–19,700 ft) away from the escarpment. The wind is thought to be named after the 'Helm' or 'helmet' of cloud. The phenomenon was intensely studied by the meteorologist, Gordon Manley, often under atrocious conditions. He interpreted the wind as a standing wave, which may occur in the lee of high ground.

Although the Helm wind from Cross Fell affects the northern end of the Eden Valley, the southern end of the valley is subject to similar conditions when air cascades over the Mallerstang escarpment. It, too, produces a loud roaring sound. Although the Helm wind is the only named British wind, similar conditions do occur elsewhere in the lee of mountainous ground.

Helm wind

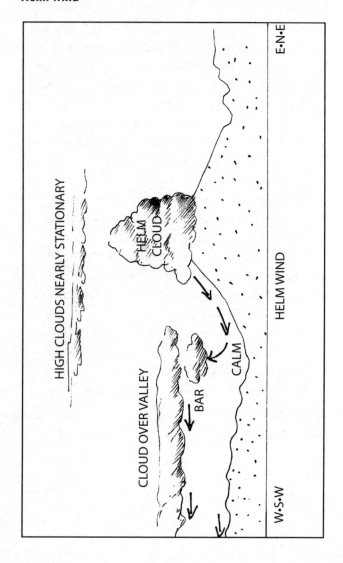

'My dear Rick,' said Mr Jarndyce, poking the fire; 'I'll take an oath that it's either in the east, or going to be. I am always conscious of an uncomfortable sensation now and then when the wind is blowing in the east.'

Charles Dickens, *Bleak House*

HIGHS AND LOWS

Windiest place

The windiest place in the British Isles is Tiree, the westernmost island of the Inner Hebrides, which has a mean annual wind speed of 14.6 knots (27 kph).

Greatest gust at a low-level site

On 13 February 1989, a wind-speed of 123 knots (228 kph) was recorded at Fraserburgh in Aberdeenshire.

Highest gust speed at a high-level site

A gust of 150 knots (278 kph) was recorded at the Cairngorm Summit on 20 March 1986.

Highest recorded speed of the jet stream

On 13 December 1967, the phenomenal speed of 353 knots (656 kph) was recorded over South Uist in the Outer Hebrides. (Although there have been some doubts about this velocity, if correct it is also a world record.)

HIGHEST GUST SPEED RECORDS – BY COUNTRY (LOW-LEVEL SITES)

COUNTRY	SPEED	DATE	LOCATION
Scotland	123 knots (229 kph)	13 Feb. 1989	Fraserburgh, Aberdeenshire
Northern Ireland	108 knots (200 kph)	12 Jan. 1974	Kilkeel, County Down
Wales	108 knots (200 kph)	28 Oct. 1989	Rhoose, Vale of Glamorgan
England	103 knots (190 kph)	15 Dec. 1979	Gwennap Head, Cornwall

HIGHEST GUST SPEED RECORDS – BY DISTRICT (LOW-LEVEL SITES)

DISTRICT	SPEED	DATE	LOCATION
Scotland E	123 knots (229 kph)	13 Feb. 1989	Fraserburgh, Aberdeenshire
Scotland N	118 knots (219 kph)	7 Feb. 1969	Kirkwall, Orkney

DISTRICT	SPEED	DATE	LOCATION
Wales S	108 knots (200 kph)	28 Oct. 1989	Rhoose, Vale of Glamorgan
England SW	103 knots (190 kph)	15 Dec. 1979	Gwennap Head, Cornwall
England SE & Central S	100 knots (185 kph)	4 Jan. 1998	Needles Old Battery, Isle of Wight
England SE & Central S	100 knots (185 kph)	16 Oct. 1987	Shoreham-by-Sea, West Sussex
Wales N	97 knots (180 kph)	24 Dec. 1997	Aberdaron, Gwynedd
England E & NE	95 knots (175 kph)	2 June 1975	South Gare, North Yorkshire
Midlands	91 knots (169 kph)	2 Jan. 1976	Wittering, Cambridgeshire
England NW	88 knots (163 kph)	13 Jan. 1984	Sellafield, Cumbria
England NW	88 knots (163 kph)	16 Jan. 1984	Sellafield, Cumbria
England NW	88 knots (163 kph)	8 Jan. 2005	St Bees Head, Cumbria
Scotland W	88 knots (163 kph)	5 Dec. 1972	Hunterston, Ayrshire
East Anglia	87 knots (161 kph)	16 Oct. 1987	Shoeburyness, Essex

FASCINATING FACTS:
FÖHN WINDS

The Helm wind, the only British named wind, is technically known as a föhn wind or rain-shadow wind. Originally recognised and named for the southerly wind that crosses the Alps, föhn winds tend to be warm and dry. This is because as they rise over the mountains they cool and deposit much of their moisture as rain or snow. When they descend beyond the mountains they warm at a greater rate because they have lost their moisture. In Britain, for example, the prevailing south-westerly winds rise over the Cairngorm and Grampian Mountains in Scotland, lose their moisture as rain or snow and descend, warmer and drier, on the eastern side. This region of north-east Scotland has relatively warm temperatures and is drier than a corresponding area on the western side. Similarly, the reverse effect occurs if the winds are from the east, when the area to the west of the Cairngorms and Grampians becomes warmer. In the case of the Helm wind, the air above the Pennines around Cross Fell is extremely cold, so the Helm itself over the Eden Valley is not only strong but still feels cold and blustery.

> I've lived in good climate,
> and it bores the hell out of me. I like
> weather rather than climate.
>
> John Steinbeck

WEATHER WORDS

Blenter (Scots): Gusty wind.

Blirtie/Blirty (Scots): A rarely used word nowadays for blustery weather.

Bomb: This is a respectable meteorological term with a specific meaning. It describes the sudden, dramatic deepening of a depression, where the central pressure drops by at least 1 millibar per hour for a whole day (24 hours). Such bombs are usually accompanied by extremely high winds.

Bub (Scots): Squall.

Chwefror a chwyth y neidr oddiar ei nyth (Old Welsh): 'February wind blows the snake off its nest.'

Flaff (Scots): Blow in gusts.

Gaoth (Irish): Wind.

Gwynt ffroen yr ych (Old Welsh): East wind, literally 'the wind of the ox's nostril'.

Gwynt traed y meirw (Old Welsh): This describes a cold east wind, literally the wind of dead men's feet, referring to the custom of burying the dead with their feet pointing eastwards.

Gwynt y creigiau (Old Welsh): North-west wind, the wind of Snowdonia rocks.

Feothan (Irish): Breeze.

Flan (Scots): A sudden strong gust of wind. The isolated island of Foula in the Shetlands is noted for the dramatic flans that it experiences when the wind is in a certain direction, and caused by the island's rugged nature.

Hushle (Scots): Strong wind.

Mae'n wyntog (Welsh): It's windy.

Pipple (Tudor/Stuart): The murmuring sound of a gentle wind.

Pirl (Scots): Gentle breeze.

Reevin (Scots): Gusty.

Séideán ghaoithe (Irish): Blast (of wind).

Snell (Scots): A bitingly cold wind.

Sweevil (Scots): Gust of wind.

Tattery (Scots): Very windy.

Tirl (Scots): Breeze.

Tousie (Scots): Blustery.

Uitwaaien (Dutch): To walk in windy weather for fun.

Williwaw (English): A sudden blast of wind descending from a mountainous coast to the sea.

Yowdendrift (Scots): Snow driven by wind.

The sharper the blast
The sooner 'tis past.

Charles Wesley

FASCINATING FACTS:
HOW TO WIN A BALLOON RACE

Balloon races, and the release of balloons at significant events, are now frowned upon, because of the hazard balloons pose to wildlife (apart from making litter). Amongst other problems, cows swallow them when they come down on land, and marine turtles try to eat them thinking they are jellyfish. However, the best way to stand a chance of winning a balloon race if you do take part in one, is to make sure that your balloon is not fully inflated – only just enough to cause it to lift off the ground. Fully inflated balloons may rise faster, but they soon burst when they expand with the lower pressure at height. Balloons with less lift are carried farther away from the release point by the wind before they rise too high and burst.

CHAPTER 5

STORMS

A horrid stillness first invades the ear,
And in that silence we the tempest hear.

John Dryden

WEATHER PEOPLE

Henry Winstanley (1644–1703): Unfortunately his wish was granted

The first lighthouse on the Eddystone rocks, a major hazard outside the port of Plymouth, was constructed by Henry Winstanley, who was a painter and engineer. He later became a merchant, but two of his ships were wrecked on the Eddystone rocks. Being told it was too hazardous to construct a lighthouse there, he was determined to build one himself. Construction began in 1696, but in 1697 Britain was at war with France and, the Admiralty having ordered the ship protecting the workers away, a French privateer destroyed the lighthouse's foundations and carried Winstanley to France. He was released on the orders of the French King, Louis XIV, who famously said 'France is at war with England, not with humanity'. The first lighthouse was completed in 1698, but suffered damage in the following winter. Winstanley rebuilt it, higher, with more stonework and greater decoration. No ships were wrecked on the Eddystone rocks during the five years both lighthouses were in operation.

Winstanley apparently said that he would like to be present on the Eddystone in 'the greatest storm there ever was'. He had his wish. He was present, to supervise repairs, in the greatest storm that Britain has ever experienced (26–27 November 1703, OS, 7–8 December, NS); the lighthouse was completely destroyed and Winstanley lost his life.

FASCINATING FACTS:
SPRITES, ELVES AND TROLLS

Sprites are legendary, supernatural creatures, often associated with water. Anyone who has read Tolkien's *Lord of the Rings*, Rowling's *Harry Potter* books, or any Scandinavian legends will have heard of elves and trolls. But all three words for these mythical beings are respectable meteorological terms for optical phenomena associated with thunderstorms.

Sprites

Sprites were the first to be discovered, only being confirmed in 1989. They are short-lived, weak, luminous phenomena that occur high above decaying thunderstorms at altitudes of between 50–80 km (around 160,000 and 260,000 ft). They last just a few milliseconds and look somewhat like jellyfish or carrots, with a reddish main body, faint tops (like carrot 'leaves') that reach upwards, and bluish tendrils extending downwards. They are believed to be caused by electrons accelerated upwards by lightning discharges.

Elves

Elves are even shorter-lived than sprites, lasting about one millisecond, and may be detected only with sophisticated equipment. They occur at altitudes of 85–100 km (53–62 miles), and are often associated with sprites. They start as a central point and expand (faster than the speed of light) as a ring, with diameters reaching several hundred kilometres. They occur when a powerful electromagnetic pulse, generated by a lightning discharge, reaches the ionosphere, accelerating electrons that collide with nitrogen atoms that then radiate faint red light.

Trolls

Trolls were discovered and given a name to match those of sprites and elves. Then someone had to think of a term for which 'troll' would be the acronym. Eventually they came up with 'Transient Red Optical Luminous Lineament'. These trolls are luminous blobs that race upwards from the tops of thunderclouds, leaving a faint trail behind them and peter out at altitudes of around 50 km (164,000 ft). They seem to occur after particularly energetic sprites and resemble another type of phenomenon known as 'blue jets', but are red.

> Poor naked wretches, wheresoe'er
> you are, that bide the pelting
> of this pitiless storm.
>
> William Shakespeare, *King Lear*

NOTABLE WEATHER EVENTS

The Armada storms, 1588

Although the defeat of the Spanish Armada in 1588 is often credited to the successful efforts of Queen Elizabeth's naval forces, the weather during the period played an extremely important part. The basic aim of the Armada was to allow forces under the Duke of Parma to invade England from the then Spanish Netherlands. Initial actions off Plymouth and the Isle of Wight were inconclusive and the Spanish fleet anchored off Calais. On the night of 28 August, the English sent in fireships, causing many of the Spanish ships to cut their anchor cables to escape. This was to prove of great significance much later. The most successful English naval action occurred off Gravelines, part of Flanders, the nearest part of the Spanish Netherlands to England. Five Spanish ships were lost. A strong southerly gale then arose, driving the Spanish ships away from the shore and causing the rendezvous with the army under the Duke of Parma to be abandoned. The severe winds drove the Spanish ships up the North Sea, during which time they were harried by the English fleet, which ceased pursuit in the area of the Firth of Forth. The only course available to the Armada was to round the north of Scotland and make south for Spain. They did so, but were unaware that the North Atlantic Drift was setting them farther to the east than they calculated. An extremely deep depression with its accompanying severe gale-force winds then struck the fleet, driving them eastwards and on to the lee shores of the Hebrides and Ireland. Because of the loss of their anchors, most of the fleet was unable to avoid being wrecked. More ships and sailors were lost to the extreme weather than to naval action.

Ironically, after the unsuccessful expedition against Spain in the following year, 1589, known as the 'English Armada', 'the Counter Armada' and 'the Drake–Norris Expedition', 14 English ships were lost to Spanish naval action, but as many as 40 were wrecked by storms as the fleet tried to return to England.

When rain comes before wind,
Halyards, sheets and braces mind;
But – When wind comes before rain,
Soon you may make sail again.

Robert FitzRoy

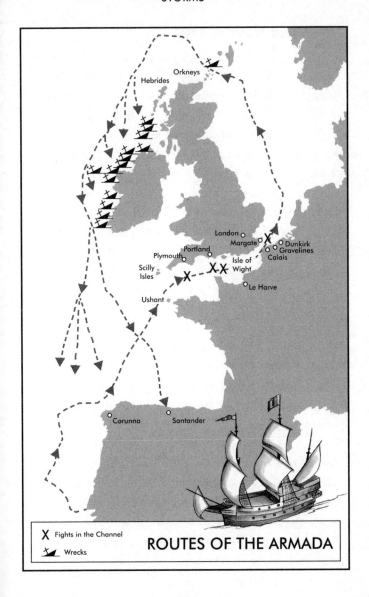

X Fights in the Channel

Wrecks

ROUTES OF THE ARMADA

Orkneys

Hebrides

London

Margate

Dunkirk

Gravelines

Calais

Portland

Plymouth

Isle of Wight

Scilly
Isles

Le Harve

Ushant

Corunna

Santander

The Great Storm of 1703

The greatest storm in historical times occurred on 26 November 1703 (OS) or 7 December 1703 (NS). It was particularly severe in the southern half of Britain. Although records are fragmentary at that period, the only comparable storm (from a meteorological point of view) is the event of 16–17 October 1987. The storm of 1703, however, was notable, not only for the widespread damage and destruction on land but for the extreme loss of life and shipping at sea. The Royal Navy suffered great losses, including the whole of the Channel Fleet. Merchant shipping was particularly badly hit, and no less than some 700 ships were swept from their moorings in the Pool of London and hurled together in a single mass, blocking the Thames. There have been various estimates of the overall loss of life, ranging from 8,000 to 15,000. The sustained high winds, estimated to have reached about 70 knots (130 kph) caused considerable destruction to standing timber, roofs, chimneys and (in particular) to windmills, over 400 of which were destroyed. Some 2,000 chimneys were blown down in London alone, where Queen Anne was forced to take shelter in a cellar at St James's Palace, when chimneys and part of the roof collapsed. There was widespread flooding of areas such as the Somerset Levels, where some hundreds of people were drowned, with a major loss of livestock. The lowest recorded pressure (in Essex) was 14.1 inches of mercury (973 millibars), but it has been suggested that the storm's central pressure, over the Midlands, may have been as low as 950 millibars (comparable to, or perhaps slightly less than, the lowest recorded pressure of 957 millibars for the 1987 storm and 953 millibars for the Burns' Day Storm).

Accounts of the 1703 disaster are viewed by some as the beginning of modern-style journalism. Daniel Defoe made a particular effort to collect and collate information about the effects of the storm, and his work *The Storm*, published in 1704, is the major source of detailed information about the event. It is also the first comprehensive account of a notable natural event to be published.

> … a calamity so Dreadful and Astonishing, that the like hath not been Seen or Felt, in the Memory of any Person Living in this Our Kingdom.
>
> From the public proclamation issued by Queen Anne, following the Great Storm of 1703

The Royal Charter *Storm, 1859*

What is possibly the most extreme storm in the Irish Sea in the nineteenth century occurred on 25–26 October 1859. A depression, with an extremely low central pressure, brought extensive damage to Devon and Cornwall, but appears to have intensified as it moved northwards, affecting the whole of Wales, before eventually crossing Scotland. Winds of Force 12 on the Beaufort Scale were experienced with speeds believed to have been well in excess of 87 knots (160 kph). As many as 133 ships were sunk by the storm and about 90 severely damaged, and many sailors lost their lives. The worst individual disaster was the loss of the steam clipper *Royal Charter*, which, on the final leg of its voyage from Melbourne to Liverpool, was driven ashore on the coast of Anglesey in the early morning of 26 October. More than 450 people were killed by this one incident. The storm caused some deaths onshore and overall about 800 individuals lost their lives.

The loss of the *Royal Charter* had a great effect on the general public, but was of perhaps even greater significance meteorologically, because Vice-Admiral Robert FitzRoy, then Meteorological Statist to the Board of Trade, made an analysis of the storm, which then induced him to introduce methods of predicting the weather, which he described as 'weather forecasting', and also to introduce storm cones at ports to warn sailors of imminent gales and storms.

The Derby Day Disaster, 1911

The weather on the day that the Derby is held at Epsom Downs (generally in the latter half of May) has often been extremely varied, illustrating how fickle the weather may be during that part of the year. On the actual race day there has been torrential rain, blistering heat and exceptionally cold conditions with freezing winds, frost, ice and driving snow. The worst disaster occurred on 31 May 1911. A generally slack pressure situation initially gave plenty of sunshine over southern England, and these conditions continued into the afternoon, proving ideal for the creation of heavy thunderstoms, several of which were extremely severe. The worst of these affected the Epsom area of Surrey just as racing was finished at around 17.00 GMT. The skies darkened and torrential rain began to fall. At nearby Banstead, 62 mm (2.44 in) of rain fell in just 50 minutes (91.2 mm (3.59 in) fell over the 24-hour period). Large hail between 37–50 mm (1.5–2 in) in diameter fell at Sutton, also in Surrey. The electrical activity was phenomenal: no less than 159 lightning strokes were counted at Epsom in one 15-minute period between 17.30 and 17.45 GMT. Conditions for people leaving the racetrack were atrocious, with torrential rain and hail 'the size of walnuts', almost continuous lightning and thunder, water-logged ground and mud. There is some disagreement about the number of people killed by lightning, probably because there were casualties from other thunderstorms in southern England that day – although none of the storms were as severe as the one at Epsom. Three people were killed in the immediate vicinity of the racecourse and a large number injured. Three haystacks were set on fire, and there were landslips that blocked the railway at nearby Merstham and Coulsdon, as well as severe flooding of the railway at Epsom itself. Overall the death toll for southern England that day is most often put at 17, together with four horses (not racehorses) killed by lightning.

The Hampstead Storm, 1975

The heaviest rainfall recorded in London – indeed, in any British urban area – occurred on 14 August 1975. Intense thunderstorms developed across eastern England and one very heavy storm affected north-west London, causing serious flooding. Hampstead Climatological Station recorded 170.8 mm (6.72 in) of rain that day, with most of the rain falling between 17.30 and 20.00 GMT. Several other observing stations in that part of London recorded more than 100 mm (3.94 in) during that observing day.

The Fastnet Race Storm, 1979

In 1979, the extreme weather turned the Fastnet Race into a disaster. The race is a major event in the yacht-racing calendar, held every two years. In 1979 it was also part of the Admiral's Cup series. The race is over 964 km (605 miles) from Cowes in the Isle of Wight to the Fastnet Rock off southern Ireland, ending at Plymouth after passing south of the Isles of Scilly.

Weather for the 1979 race, which started on 11 August, was predicted to have south-westerly winds of Force 4 to 5, increasing to Force 6 or 7. However, a strong depression formed over the Atlantic on 11–12 August, which rapidly deepened on 13 August, and deviated from its original track, turning towards the north-east. By 14 August it was centred over Wexford in southern Ireland with extreme winds on its southern flank over the sea (and the yachts taking part in the race). The central pressure in the depression dropped to 979 hectopascals and Meteorological Office observations and calculations suggest that the wind over the sea reached Force 10, although some of the competitors believed it reached Force 11. The winds and extreme 'mountainous' seas took a great toll of the yachts in the race. Of 306 entries, five yachts sank, 100 were knocked down and no less than 77 rolled upside-down (known as 'turtling') at least once during the storm. The largest ever rescue effort known in peace time was launched, involving Royal Navy, Irish, United States and Dutch Navy ships, six lifeboats, eight Royal Navy helicopters, seven RAF helicopters, two Irish Air Corps aircraft, two RAF Nimrod aircraft, civilian tugs, trawlers and tankers. Around 4,000 rescuers were involved. Some 136 yachtsmen were rescued but 15 competitors and three rescuers lost their lives. Only 86 yachts finished the race, 194 being forced to retire and 24 yachts being abandoned (with five believed to have sunk).

On land there were severe gales in south-western England and west Wales, with a gust of 65 knots (121 kph) recorded at Milford Haven in Pembrokeshire. There was also damage in affected areas on land where four people were killed during the storm.

A somewhat similar disaster hit the Sydney to Hobart yacht race in 1998, when another storm, which also had near-hurricane-force winds, caused five yachts to sink and six people to lose their lives. Of 115 yachts that started the race, only 44 reached Hobart.

Hail, coke, stones and a crab from the sky, 1983

On 5 June 1983 a series of violent thunderstorms swept eastwards from Lyme Bay across Weymouth, Poole and Bournemouth, and on into Hampshire and Sussex. They were accompanied by torrential rain (74 mm (2.91 in) fell at Winfrith in Dorset) and heavy hail with stones the size of golf balls (43 mm/1.69 in) in the Poole and Bournemouth areas. One stone was measured at 65 mm (2.56 in) across. The earliest system that day had associated tornadic activity that raised material from the ground, possibly from Hamworthy, Poole, because large amounts of coke fell at Poole and Bournemouth, together with some stones and a piece of coal. Farther east, a crab fell from one storm over Sussex. Some hailstones were found that had formed around what appeared to be roof chippings. The storms probably originated as supercell systems that had formed over France and then crossed the English Channel.

The Great Storm of 1987

Although the storm that hit southern England on 16 October 1987 is often compared with the Great Storm of 1703, in reality, storms of this ferocity are not that uncommon over the British Isles. Normally the centres of such storms pass to the north of the British Isles, and the extreme winds affect the north of Scotland. What was unusual about the October 1987 storm was the fact that its main impact was felt over south-eastern England. A central pressure of just 957 millibars was recorded at Exeter at 02.00 GMT and one of 951 millibars over the English Channel. The centre of the depression (known technically and perhaps more dramatically as an extratropical cyclone) passed from southern Devon, across the Midlands to the Humber Estuary. The extreme winds occurred on its southern flank, with gusts of 90 knots (167 kph) at many locations along the south coast and one recorded gust of 100 knots (185 kph) at Shoreham-by-Sea in West Sussex. (A higher gust of 106 knots (196 kph) has also been reported for Gorleston in Norfolk.) The highest hourly mean wind speed recorded was 75 knots (139 kph) at the Royal Sovereign Lighthouse in the English Channel (south-east of Beachy Head in East Sussex).

It is now known that the extreme winds were produced by a 'Sting Jet', a phenomenon that was only discovered subsequently after analysis of the data recorded during the storm. It should be noted that this storm was not 'a hurricane' as widely reported in the media, but that the winds were of 'hurricane force'. (Hurricanes cannot occur over Britain.)

The storm caused widespread damage, and major problems with the electrical grid over southern England, which eventually had to be shut down to prevent a catastrophic failure. It was particularly destructive of trees, because many were still in full leaf and the ground was waterlogged from earlier heavy rains. It is estimated that 15 million trees were blown down. Luckily the worst winds occurred during the night, between 03.00 and 07.00 GMT when few people were outdoors. Although 19 people were killed, the death toll would undoubtedly have been much higher if the extreme winds had struck during the day.

Subsequently, the Meteorological Office was subjected to considerable criticism for its apparent failure to forecast the extreme conditions. In fact, early forecasts, some days beforehand, did suggest the likelihood of extreme winds over southern Britain, but later revisions predicted that the worst of the winds would occur farther south, over France. (In the event, this was true, because the devastation in France was even greater than in England, although fewer individuals were killed.) Later analysis showed that the forecasts were incomplete because of a lack of observations from out in the Atlantic, and there is also the suggestion that industrial action prevented data from French stations being passed to the Meteorological Office. The Meteorological Office implemented a number of changes and improvements as a result of their investigations, including alterations to the training of forecasters and modifications to the numerical-forecasting models. It established the National Severe Weather Warning Service. Subsequently improvements were made to increase data coverage over the Atlantic, including automatic buoys, and increased data acquisition from ships and aircraft.

The Burns' Day Storm, 1990

On 25 January 1990, slightly more than two years after the extreme storm of 16 October 1987, another severe event struck Britain. The centre of this deep depression (central pressure 953 millibars, that is, slightly lower than in the 1987 storm) tracked across southern Scotland, crossing the home of Robert Burns in Ayrshire on the anniversary of his birth, causing the event to become known as 'the Burns' Day Storm'. Once again, the most extreme winds were to the south of the depression's centre, with great damage across much of both Wales and England. In many cases, the winds were even higher than on the previous occasion and they also affected a far wider area. Because the worst winds struck during the day, the death toll was greater than in the 1987 October Storm, with more than twice as many (47) people being killed in Britain. Mean wind speeds of 40–50 knots (74–93 kph) were widespread, with speeds of 60–65 knots (111–120 kph) over exposed coasts of southern and western Wales. The highest gust was 93 knots (172 kph) at Aberporth in Ceredigon. Considerable destruction of buildings took place right across the country, from the Scilly Isles in the south-west to the east coast of England.

(The storm actually continued across the North Sea and caused damage and a comparable number of deaths in Denmark.) Numerous record hourly mean and gust speeds were recorded for individual stations during the event and many records remain unbroken to this day.

There was widespread disruption to road and rail traffic, and interruption of power supplies, but the number of trees uprooted was less than in the storm of October 1987, probably because, given the time of year, most deciduous trees were leafless.

On this occasion, the changes implemented by the Meteorological Office after the 1987 storm were effective and led to adequate warnings of the impending storm being issued some days in advance. There were accurate predictions of the severity of the winds, the timing of their onset, and the general location and motion of the centre of the low.

The Braer Storm, 1993

On 8 January 1993 a weak frontal wave started to develop into a depression. It began moving north-east but then amalgamated with another low-pressure area, originally to its south-east. The deepening was enhanced by a strong sea-surface temperature gradient along its path and by a very strong jet stream, which had winds of some 440 kph (270 mph). The system became a 'bomb', deepening by more than 1 millibar per hour for over 24 hours. By 10 January it had become the deepest extratropical cyclone ever recorded in the

North Atlantic, with a central pressure estimated at 916 millibars. The winds associated with this depression were gale-force across the whole of the North Atlantic, and on its southern side they reached hurricane force. At its deepest, the centre was located north-west of Scotland. The ocean weather ship *Cumulus*, which was stationed at latitude 57° 05' N, and longitude 16° 18' W (south of the centre), recorded a maximum gust of 105 knots (194 kph), and a similar value was recorded at the very remote island of North Rona, which lies 44 km (27 miles) north-north-east of the Butt of Lewis, the northernmost point of Lewis in the Outer Hebrides. Gusts of 104 knots (190 kph) were measured in the north-west of the Scottish mainland.

On this occasion, the Shipping Forecast that was broadcast by the BBC gave the most extreme forecast ever issued, which was completely without precedent: 'Rockall, Malin, Hebrides, Bailey. South-west hurricane Force 12 or more.' (Note the use of the words 'or more'.)

Nearly a week before, on 5 January, the MV *Braer* tanker had reported loss of engine power. She was then south of Sumburgh Head, the southernmost point of Mainland, the principal island in the Shetlands. Driven by the winds of a smaller, earlier depression, the vessel, despite efforts to bring her under control with a tow, had grounded near the southern tip of Mainland and started to leak her cargo of light crude oil. When, a few days later, the vessel was hit by the force of the wind and waves of the exceptional depression, it broke up completely, spilling its whole cargo of oil into the sea. Although it was a light crude oil, it still caused an environmental disaster, with thousands of sea birds being killed. It is primarily for this reason that the storm has come to be known as 'the Braer Storm'. However, there was one redeeming feature: the extreme waves and winds rapidly dispersed the oil, preventing an even worse disaster.

North Sea Storm Surge, 2007

Following the disastrous floods of 1953, the Meteorological Office set up special arrangements to predict storm surges that might pose major threats. The weather during the winter period caused the worst problems. In early November 2007, predictions indicated that very severe conditions were likely to occur on 9 November. All the indications were that the situation would be the worst since 1953. A storm surge of 3 m (10 ft) above normal levels was predicted. In the event, conditions were not quite as bad as expected, primarily because the winds tended to be blowing offshore – and thus reducing wave heights along the coast – rather than onshore as on the earlier occasion. Nevertheless, there was some flooding and damage, but no loss of life. At Great Yarmouth, the sea came within just 10 cm (4 in) of overtopping the sea defences. The Thames Barrier was closed twice that day, before the times of high tide.

Winter storms, 2013–14

On 5 December 2013, storm-force winds affected Scotland and northern England, with extensive power cuts, two fatalities and major disruption to transport – the whole of Scotland's rail network was shut down, and all flights were affected. From 5–7 December, the depression responsible deepened rapidly and tracked from north of Scotland to the Baltic, producing a major storm surge in the North Sea. Levels were predicted to be even worse than those that caused the disaster in 1953. Thousands of people on the east coast and in Kent were advised to evacuate their homes. Fortunately, the majority of the sea defences held, although some homes were flooded in Hull, and there was flooding of the town centre in Boston, Lincolnshire. At Hemsby in Norfolk, a lifeboat station was washed away and seven homes were lost when cliffs collapsed. Jaywick in Essex was flooded and, on the other side of the country, Rhyll in Denbighshire was also flooded as a result of a separate storm surge, resulting from the gale-force winds and high tides. The Thames Barrier was closed and prevented flooding of low-lying areas of London. Two people were killed by the accompanying high winds.

The extreme conditions in the North Sea in December had an unexpected bonus effect. A remarkable, previously unknown, submerged forest was uncovered in the area known as Doggerland, which, at the time of lower sea levels, was dry land stretching between Britain and Germany, once populated by hunter-gatherers.

Another severe storm hit Northern Ireland and western Scotland on 18–19 December, with travel disrupted, power cuts and a certain amount of damage. On 23–24 December, the south coast of England was hit by a severe storm, resulting in widespread flooding from Dorset to Kent, as well as power cuts. Some 50,000 homes were without power over Christmas. There was one fatality in Devon, rail services were disrupted and Gatwick airport was flooded. More severe weather occurred on 26–27 December and 30–31 December. This time it was Wales and southern Scotland that suffered the brunt of the storm. In early January there was widespread flooding by the rivers Severn and Thames. The Somerset Levels were also flooded. High tides and large waves caused major damage around Wales (the promenade at Aberystwyth was largely destroyed) and along the south coast, with one death in Devon. The storms in December 2013 and early January 2014 caused seven fatalities in England, with some 1,700 homes flooded.

Winter storm, 2014

The severe weather of December 2013 and early January 2014 continued into late January and February, to such an extent that the overall period saw more destructive storms than any winter for more than 20 years. Flooding continued to affect areas around the Severn and Thames, and the Somerset Levels, in particular, remained flooded for weeks from the end of

December, with several villages only accessible by boat. The flooding there also caused major problems on the main railway line to the south-west.

The succession of major storms during this period, together with high tides, caused major tidal surges and enormous waves. The energy in the waves created exceptional damage to the coast in south-western England and West Wales, and completely remodelled the beaches in numerous places. (Photographs show the waves breaking against the coast and reaching phenomenal heights.) On 4–5 February, the sea wall at Dawlish was destroyed, severing the main rail link to Plymouth and Cornwall for several weeks. The same storm damaged sea defences and created coastal flooding in Cornwall, Devon and Dorset.

The most powerful storm occurred on 12 February, when wave heights of 25 m (82 ft) and hurricane-force winds were recorded by a gas platform off southern Ireland. A red warning (the highest category) for winds was issued by the Met Office for North Wales and north-western England. That particular storm resulted in about 100,000 homes and businesses being without power. A number of people were killed by falling trees, broken power lines and collapsing masonry. Severe flood warnings were issued for the south coast, the Somerset Levels and the Thames Valley, and flooding did occur in those areas.

FASCINATING FACTS:
THE VARIOUS TYPES OF
LIGHTNING

People often think of lightning as two (or at most three) different types: fork lightning, sheet lightning and 'heat' lightning. But these divisions are not actually based in reality. Fork lightning is merely lightning where one can see the actual lightning channel. Sheet lightning, by contrast, is where the lightning stroke itself is hidden (usually by cloud) and only a flash is seen. The term 'heat lightning' is sometimes used when no thunder is heard.

Meteorologists do recognise three forms of lightning, but these are based on completely different features. These three are: cloud-to-cloud (CC) lightning, sometimes called 'intercloud lightning'; cloud-to-ground (CG) lightning; and lightning between different parts of the same cloud (IC lightning), sometimes known as 'intra-cloud lightning'. These terms really speak for themselves. With the intra-cloud lightning the path of the stroke is usually hidden well within the cloud, so this tends to be the form that is given the popular term 'sheet lightning'.

MYTHS AND MISTAKES

The 'hurricane' that never existed

Legend has it that Michael Fish, giving an evening weather forecast on 15 October 1987, said, 'A lady has phoned to ask if there is going to be a hurricane tonight... there is not!' But that's not what happened at all. In a morning forecast, where he warned of extreme winds in Britain, the additional comment that he made was about whether a hurricane over the Caribbean would affect Florida later. Fish was not even on duty to give the evening forecast, which was actually given by Bill Giles, who warned of high winds and heavy rain along the Channel coasts. In fact, the Meteorological Office forecasters were hampered in producing a full forecast by industrial action in France, which prevented them from obtaining vital data from French meteorological stations. In addition, there was a lack of data from out in the Atlantic, such data being essential in making accurate predictions of how storms are likely to develop.

In any case, the storm of 15–16 October 1989 was not a 'hurricane' – a term which has a very specific meteorological meaning – but a deep depression, accompanied by 'hurricane-force' winds.

FASCINATING FACTS:
HOW FAR AWAY IS THE
LIGHTNING?

The approximate distance of any lightning stroke may be estimated by counting the seconds between seeing the flash and hearing the thunder. Sound travels very much slower than light (about 340 m per second as against 300,000,000 m per second). A delay of three seconds corresponds to about 1 km (five seconds to about 1 mile). Thunder is not normally heard more than about 20 km (12.5 miles) away (so the delay would be about 60 seconds – one minute). Beyond that distance, the flash may be seen but no thunder is heard – this is when people often say they have seen 'heat lightning'. On very rare occasions, and under very unusual atmospheric conditions, thunder has been heard over a distance of more than 60 km (37 miles).

WEATHER LORE

'As the days grow longer,
The storms grow stronger.
As the day lengthens,
The cold strengthens.'

Unfortunately, this piece of 'lore', which originated in Yorkshire, has little meteorological basis. The worst storms tend to occur in winter (when the days are shorter). There is perhaps one sense in which the second part has a slight tinge of accuracy. Days begin to lengthen after the winter solstice in December, and the coldest part of the year usually comes in January and early February.

FASCINATING FACTS:
THE CAUSES OF LIGHTNING

Lightning has been studied for many years and meteorologists have a good idea of how the actual strokes propagate through the air, because the paths themselves may be observed and studied in detail. They take place in several stages, which may be briefly described as follows: An initial 'leader' makes its way down from the cloud towards the ground in stages and in a zigzag fashion, with multiple branches. When one of these makes contact with streamers that have propagated upwards, the main discharge current (the 'return stroke') passes up from the ground to the cloud. There may be several, separate 're-strikes', each preceded by a downward-moving dart leader. (As many as 30 separate strokes have been known to occur, but the number is usually far smaller.) It is still not fully understood how the initial leader strokes are formed.

But the initial process by which the electrical charges are built up within a cumulonimbus cloud are still poorly understood. The processes involved are hidden within the cloud and are difficult to study or simulate in the laboratory. It would seem that freezing ('glaciation') is essential. The separation into positive and negative charges appears to occur when ice particles and water droplets collide as a result of the great turbulence within the cloud. The ice becomes positively charged and the surrounding mixture of snow and water droplets negatively charged. The powerful turbulence and updraughts separate the lighter ice particles and carry them upwards. The top of the cumulonimbus cloud becomes positively charged, and its base acquires a negative charge.

A positive charge builds up on the ground below the negatively charged cloud-base and follows it across the landscape as the cloud moves with the wind. That positive charge 'climbs' over high obstacles on the ground, such as electricity pylons, trees and buildings and, as it does so, obviously reduces the gap between cloud and the surface. When a sufficiently high 'space charge' has built up, a lightning stroke occurs, following the shortest path, which is why the highest objects are the ones that are usually hit.

HIGHS AND LOWS

Lowest recorded mainland pressure

The lowest pressure ever recorded at a mainland, low-level site was 925.5 millibars at Ochtertyre in Perthshire on 26 January 1884.

Lowest pressure

The lowest sea-level pressure at any location in the British Isles was recorded on 10 January 1993 at the remote island of North Rona, off north-west Scotland. The pressure of 916 millibars occurred on the same day as the station experienced winds of 105 knots (194 kph). This extremely low pressure occurred in conjunction with the storm that has come to be called 'the Braer Storm', described earlier.

 FASCINATING FACTS:
LIGHTNING SAFETY

If an active thunderstorm appears to be heading towards you, it is wise to take precautions. It is safer to be indoors than outside, and sitting inside a car is normally quite safe, provided the bodywork is metal and not fibreglass. A metal body forms what is known as a Faraday cage, and will conduct any discharge safely around any occupants. Inside a house, it is wise to stay away from windows, electrical wiring and plumbing. A lightning strike normally takes a path to earth through the electrical wiring or the water-filled pipes. Telephones and any electrical appliances are at risk and ideally should be unplugged, just as television aerials should be disconnected.

If trapped outside, it is probably common knowledge by now that it is particularly dangerous to stand under an umbrella or an isolated tree. (A dry ditch is relatively safe, but not one in which there is water. Water can conduct an electrical charge for long distances horizontally.) What is not well known is that although one wants to get as low as possible, so that one is not the highest object around, it is potentially dangerous to lie flat on the ground. There may be a considerable difference in electrical charge between your head and your feet, which may lead to being struck. This is why large animals such as cows and horses are sometimes killed by lightning – simply because their legs are wide apart. It is safer to crouch down, if possible on the balls of your feet, with your arms around your knees. It may sound paradoxical, but if your clothes are wet they can help to channel the charge away from the centre of your body. (The greatest danger is that the intense electrical shock will affect your heart.) In a crouched position, people have survived direct lightning strikes, because the charge has passed down their backs and jumped straight to the ground, often leaving fern-like patterns on their skin, which gradually fade away, but without any major burns.

From these three Heads we are brought down directly to speak of the Particular Storm before us; viz. The Greatest, the Longest in Duration, the widest in Extent, of all the Tempests and Storms that History gives any Account of since the Beginning of Time.

Daniel Defoe, *The Storm*

WEATHER WORDS

Artery (Scots): Stormy.

Blirty (Scots): Changeable weather.

Coorse (Scots): Stormy.

Lichtnin (Scots): Lightning.

Mae hi'n stormus (Welsh): It's stormy.

Mor dywyll a bol buwch (Old Welsh): 'As dark [the sky] as a cow's stomach.'

Roukie (Scots): Muggy.

Shucky (Sussex dialect): Unsettled weather.

Stoirm (Irish): Storm.

Storm Fai (Old Welsh): May storm; unseasonably cold weather.

Stoirmeach (Irish): Stormy.

Taranau yn Rhagfyr arwyddant dywydd teg (Old Welsh): Thunder in December signals fine weather.

Thunder-plump (Scots): Sudden thunder shower.

'A BOLT FROM THE BLUE'

Does lightning ever come out of a clear sky as a 'bolt from the blue'? Very occasionally, lightning may seem to strike when no cloud is overhead. These lightning strikes may travel horizontally from the top of the cloud for great distances (several kilometres) before turning towards the ground, so they seem to arrive from the clear blue sky. Because the top of a cumulonimbus cloud has a positive charge, and the lightning channel is so long, the amount of charge necessary to break down the resistance along the path is very high. Such positive strokes are much more powerful than the far more common negative strokes which propagate between the base of clouds and the ground. They may be as much as ten times as strong. They are hotter and the discharge current lasts much longer. Luckily, such positive strikes make up less than five per cent of all lightning discharges.

Why is it that showers and even storms seem to come by chance, so that many people think it quite natural to pray for rain or fine weather, though they would consider it ridiculous to ask for an eclipse by prayer?

Henri Poincaré, *Science and Method*

CHAPTER 6
SNOW AND ICE

I like the cold weather. It means you get work done.

Noam Chomsky

FASCINATING FACTS:
COLOUR IN ICE AND SNOW

Ice crystals are clear and transparent, but if any air is trapped within them, they appear white. When rain falls on to a cold surface, such as a road, it freezes into a layer of perfectly clear, transparent ice, when it is often known as 'black ice'. The more technical, meteorological term is 'glaze'. It is this type of ice that sometimes gives rise to dramatic 'ice storms', when every object becomes covered in a sheet of ice, usually causing breakdowns in electricity supplies and telephone services (bringing down overhead lines and even sturdy electricity pylons), as well as creating hazardous driving and walking conditions and breaking trees and shrubs.

If hailstones are cut open, they often show an onion-like structure of alternating clear and opaque layers. This appearance has been created by journeys up and down within a cloud. When the growing hailstone encounters a region with liquid water droplets, these spread over the surface to give a clear layer. When multiple tiny ice crystals freeze on to a hailstone, they trap air between them and produce an opaque, white layer.

In fallen snow, although the individual ice crystals are clear, they have so many tiny flat surfaces that they reflect a lot of light and so the whole layer appears white.

The two things, love and snow, that make the world look fresh again.

Charles Finch, *The Last Enchantments*

NOTABLE WEATHER EVENTS

The winter of 1947

The winter of 1947 was particularly severe, with what was probably the most extreme quantity of snow ever encountered in Britain. Although some snow fell in December 1946 and in early January 1947, this melted with unusually warm conditions later in the month. Significant snowfall began in the third week of January and snow fell somewhere in the country on every day between 22 January and 17 March. Individual snowfalls of 60 cm (24 in) were common, as were drifts of 5 m (16 ft) or more. Most minor roads became impassable and even major roads and railways were repeatedly blocked by enormous drifts. In southern England particularly bad conditions occurred in early March when a depression moved along the English Channel on 4–5 March, giving extremely heavy snowfall, which completely disrupted rail services, quite apart from blocking all the roads.

The consequences of this extreme snowfall were particularly serious, because conditions were already difficult in the country in the aftermath of World War Two. Food rationing was still in force, together with limitations on fuel. At that time the majority of houses were dependent on coal, or used gas fires for heating, central heating being essentially unknown. Supplies of coal – most of which was transported by rail – were severely affected, with major snowdrifts repeatedly blocking railway lines. This in turn affected gas supplies, because at that period gas was produced from coal. There were interruptions to the gas supply, which was particularly dangerous, because at the time 'town gas' was poisonous, unlike the methane (natural gas) that replaced it many years later.

Many families found themselves forced to restrict heating to a single room that had an open fire, and in which they spent both day and night. Supplies of food were also heavily restricted and the already meagre rations (by modern standards) had to be reduced still further. Heavy snowfall on 24–25 January heralded major power cuts that gradually increased to as much as 25 per cent of the time and lasted for up to three hours. Water supplies were also affected with even water mains becoming frozen.

Only a few areas escaped with a light covering of snow, such as parts of the south-west, western Wales and Scotland, and the Central Valley of Scotland. Elsewhere there was lying snow on 30–40 mornings, with even longer periods in parts of the Pennines and the Cairngorms and Grampians in Scotland. Only in the winter of 1962–63 did snow persist for a longer overall period.

The winter of 1962–63

Although the actual amount of snow that fell during the winter of 1962–63 was not as great as that encountered in 1947, the overall, unbroken period during which snow covered the ground was longer. It was also the coldest winter since British records began some 330 years earlier. Only the winter of 1739–40 saw similar temperatures (and these were only fractions of a degree warmer). Snow started to fall over Scotland and northern England on Christmas Day 1962, and by mid-afternoon on 26 December it had reached southern England, where it continued almost unabated until 28 December. After a short break, heavy snowfall recommenced on 29 December, accompanied by an easterly gale that swept the fine, powdery snow into great drifts across the whole country, with the exception of northern Scotland. Road and rail transport became nearly impossible, and innumerable places were completely isolated with some drifts reaching 6–7 m (20–23 ft) deep. There were further significant snowfalls in early and mid-January and a very severe one on 6–7 February. One remarkable feature of the winter was the persistence of easterly winds, which were accompanied by bright clear days and very cold nights. The persistent easterly winds meant that much of Cornwall escaped heavy snowfall, although even slightly farther east, snow was lying for as many as 60 days during that winter. Even along the Channel coast, where any snow is rare, as many as 40 days of lying snow were reported. Much of the rest of the country experienced lying snow for 60 days or even more.

The Emley Moor mast collapse, 1969

At the time of its construction in 1966, the television transmission mast at Emley Moor in West Yorkshire was, at 385 m (1,263 ft), one of the tallest structures in the world. Most of the mast consisted of a cylindrical steel tube (275 m (902 ft) tall and 2.75 m (9 ft) in diameter), carrying a lattice structure 107 m (351 ft) high and a final capping cylinder. The whole structure was stayed by guy wires. In winter, the mast and wires often became coated in ice, which commonly fell from the guy wires, to such an extent that red warning lights were installed as hazard warning and nearby roads were even closed occasionally.

On 19 March 1969, during strong winds, the whole mast collapsed under the weight of the ice on the tower and guy wires. Although one broken wire sliced through the nearby church and transmitter buildings at the base of the mast, no one was injured. A subsequent enquiry suggested that an additional factor in the collapse was an oscillation that built up in the tower under a low, sustained wind speed. Appropriate modifications to prevent oscillations were made to two similar masts elsewhere in the country and these have remained standing.

The mast was later replaced with a tapering, reinforced concrete tower the same height as the original steel cylinder, surmounted by a 55 m (180 ft) steel lattice. This gives the tower a total height of 330.4 m (1,084 ft). It is the highest freestanding structure in the UK, and a Grade II-listed building.

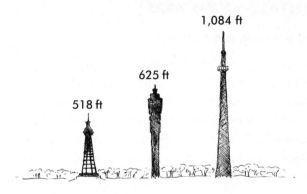

1,084 ft

625 ft

518 ft

The Emley Moor transmitter is the tallest free-standing building in the whole of the UK – higher than Blackpool Tower (left) or the BT Tower (centre).

 FASCINATING FACTS:
WATERMELON SNOW

A layer of snow on the ground does not always appear white – sometimes it may have a pink tinge. This is 'watermelon snow' – coloured just like the centre of a watermelon – and it is caused by a single-celled alga known as *Chlamydomonas nivalis*. This organism will only grow under extremely cold conditions and is normally found at altitudes of 2,500 m (8,200 ft) and above.

MYTHS AND MISTAKES

The wrong kind of snow

In Britain, railway companies, in particular, together with local authorities are often derided when they say that their problems are caused by 'the wrong kind of snow'. People say 'How can there be the wrong kind of snow? Snow is just snow.' But it's not. It may be a popular legend that the Inuit have 40 names for different kinds of snow, but to meteorologists (and railwaymen and local authorities) there are just two types of snow: wet snow and dry snow. Dry snow forms when there are low temperatures at the ground and the individual particles of ice remain separate. The snow is loose and may be easily moved by snow blowers, so it poses few problems other than the depth that may accumulate from a major snowstorm, especially if accompanied by strong winds that create deep snowdrifts. This type of snow tends to fall under very cold conditions, such as those regularly found over North America and Continental Europe.

The trouble is that in Britain we generally experience wet snow. It forms at temperatures that are only just below freezing. The real problem occurs when that snow is compressed. Under pressure it turns into water, and when the pressure is released, it refreezes into ice. (So it is the best form of snow to make snowballs, because the snowflakes freeze on to one another where they touch, when you compress them with your hands.) Under the wheels of trains or road vehicles, it liquefies, but once the pressure is released it turns into a sheet of ice. Following trains or vehicles encounter ice, rather than snow, so wheels slip and, on the roads, vehicles go into a skid. The compacted snow and ice cannot be blown aside, but requires mechanical action by snowploughs to be removed. On the railways, the moving parts of points become locked into position unless the points are provided with heaters.

FASCINATING FACTS:
SNOWFLAKES

Although the popular image of a snowflake is one of the pretty, six-sided objects that have become used everywhere at Christmastime, in fact, the six-sided form is found only in individual ice crystals. Under very cold conditions, such crystals may reach the ground, but normally multiple ice crystals clump and freeze together to produce a snowflake, which is irregular in shape, and the form in which most snow reaches the ground. Individual snowflakes have been known to reach as much as 10 cm (4 in) across.

> The winter seemed reluctant to let go its bite. It hung on cold and wet and windy long after its time.
>
> John Steinbeck, *East of Eden*

WEATHER LORE

'Plentiful berries indicate a hard winter to come.'

A surprisingly persistent common belief is that a plentiful crop of berries on holly trees, hawthorns, and hips on wild roses are nature's way of providing food for birds in the forthcoming hard winter. It is, of course, nothing of the sort. In reality, it tells you a little bit about past weather. The weather during the flowering season for the various plants, and subsequently, must have been particularly favourable for berries, hips and haws to develop.

Ice saints

In Continental Europe (but not in Britain) belief in cold 'Ice Saints' days tends to persist. The actual saints vary slightly from country to country, but usually include St Mamertus, St Pancras and St Servatius. Their feast days fall on 11 May, 12 May and 13 May, respectively. The days around this time were noted as being part of a brief spell of cold weather and usually included the last frosts. However, this was true under the old Julian calendar. After the calendar reform, the dates changed to 21–23 May. In certain countries, St Mamertus is not included, but St Boniface of Tarsus is added (14 May OS, 24 May NS). Sometimes St Sophia, nicknamed 'Cold Sophia' (15 May OS, 25 May NS) is included in the list. However, just as with the lore surrounding rain on St Swithin's day and Buchan's warm and cold spells, examination of actual weather records does not reveal any truth to these stories.

FASCINATING FACTS:
MEASURING SNOWFALL

The simplest method of determining how much snow has fallen is to use a measuring stick, plunged into the snow over a solid surface. Unfortunately, snow tends to compact under its own weight, particularly if it has been subjected to cycles of warmth – producing partial melting – and subsequent refreezing. Over an extended period, such freeze–thaw cycles cause the snow to become granular: a form known as 'firn'. Eventually, the firn in turn becomes compacted into dense glacier ice.

Another way of determining the amount of snow that has fallen is to take a vertical core, and melt the plug of snow thus obtained. In general, 30 cm (12 in) of fresh snow is approximately equal to 25 mm (0.98 in) of water.

HIGHS AND LOWS

Lowest temperature

The lowest temperature ever recorded in the British Isles was –27.2°C, which occurred on three occasions and in two different places in Scotland: at Braemar in Aberdeenshire on 10 January 1982 and on 11 February 1895. The same temperature was also recorded at Altnahara in the Highlands on 30 December 1995.

Last frost fair

The last frost fair on the Thames occurred between 1–4 February 1814, with a sudden break-up of the ice on 4 February, when a temporary alehouse was seen drifting down-river in flames.

Earliest snow in London

The earliest date on which snowfall has been recorded in London was 25 September 1895, when snow also fell at Wallington in Surrey.

Longest continuous period of snowfall

During the winter of 1947, snow fell somewhere in the country on every day between 22 January and 17 March.

Worst avalanche

The greatest loss of life from an avalanche was not in the Scottish highlands, but when an overhanging cornice of snow collapsed on 27 December 1836 at Lewes in Sussex. Eight people were killed and several houses destroyed. The event is commemorated in the name of the Snowdrop Inn.

Largest hailstone

The largest British hailstone fell at Horsham on 5 September 1958. There is disagreement about the exact size and weight. Everyone agrees that it was 55 mm (2.17 in) across, but the weights quoted vary from 140–190 g (5–7 oz). Suffice to say that the weight was more than that of a cricket ball.

FASCINATING FACTS:
'THUNDERSNOW'

During the winter of 2014–15 there was some mention in the media of 'thundersnow' over Scotland and northern England. This is simply a thunderstorm where snow reaches the ground rather than rain. Although moderately uncommon, it is not that rare and the phenomenon has been known to meteorologists for many years. It often occurs in the cold air behind the cold front of a depression, especially when the surface beneath the air is warm. This frequently happens at the beginning of winter when the sea is still relatively warm and produces vigorous convection. (Thundersnow and 'lake-effect snow' commonly occur when cold air streams across the Great Lakes in North America.) Any precipitation from clouds falls into colder air and reaches the surface as snow.

But the cumulonimbus clouds may still produce lightning and thunder. One unique feature of such clouds is that the thunder is muted ('blanketed'). Although thunder is normally heard many kilometres away – generally up to 20 km (12.5 miles) – the snow tends to muffle the sound, which may only be heard within about 3–4 km (2–2.5 miles) of the lightning.

> There's no such thing as bad weather, just soft people.
>
> Bill Bowerman

WEATHER WORDS

Egger-nogger (Sussex dialect): An uncommon term for sleet.

Fann (Scots): A snow drift.

Flag (Scots): Large snowflake.

Flichshneachta (Irish): Sleet.

Freest (Scots): Frost.

Fuair (Irish): Cold.

Gull (Scots): Cold mist.

Jeel (Scots): Extreme cold.

Leac oighir (Irish): Sheet (of ice).

Lemmed (or steeved) with the cold (Cornish): Frozen, very cold.

Mae'n bwrw eira (Welsh): It's snowing.

Mae'n oer (Welsh): It's cold.

Niwl gaeaf, arwydd eira (Old Welsh): Winter fog, sign of snow.

Peenge (Scots): Look cold and miserable.

Rhew bargod (Old Welsh): Icicle, literally, 'eaves ice'.

Shrammed (English): Completely shrivelled or numb with cold.

Síobadh sneachta (Irish): Blizzard.

Sluppra (Shetland Isles): Half-melted snow.

Sna or **Snaw** (Scots): Snow.

Snawbroo (Scots): Melted snow.

Snawwreath (Scots): Snowdrift.

Sneachta (Irish): Snow.

Snippin (Scots): Biting snow.

Stoat or **Stot** (Scots): To bounce, as may occur when heavy rain ('stottin' renn') or hailstones bounce off the ground after they have hit.

Thirling (Scots): Piercing cold.

Whummle (Scots): Avalanche.

FASCINATING FACTS:
FROST FAIRS

In the seventeenth and eighteenth centuries in particular, the River Thames froze completely from bank to bank on no less than five occasions each century. It also froze in earlier centuries, being frozen for over two months in the very severe winter of 1683–84. The last occasion was in 1814. Not only was the ice thick enough to walk and skate on, but it was sufficiently strong for people to take advantage of the situation to build booths and other temporary structures to sell hot and cold drinks and food, and offer various simple entertainments. These frost fairs, as they were known, became extremely popular and many hundreds or thousands of people might be on the ice at any one time. But the fairs were only possible because the old London Bridge seriously impeded the flow of the river. Not only were the arches very narrow, as so many were needed to span the river, but all the piers were protected by structures known as 'starlings', which were an arrangement of piles, set in a shape rather like a boat, but pointed at each end, designed to protect the piers. However, they also greatly restricted the waterway and tended to block any 'ice floes' from passing down the river, causing ice to build up upstream. When the bridge was rebuilt in 1835, with wider arches and no starlings, movement of the water (both coming downstream and tides moving upstream) was far less restricted, and heavy ice was no longer able to form. On rare occasions, the Thames may still freeze over, but this happens much farther upstream, and the ice is never very thick. (Nowadays, the Thames is tidal as far up as Teddington weir.)

I like weather. I never understand why people move someplace so that they can avoid weather.

Holly Hunter

CHAPTER 7

EXTREMES

> Weather can kill you so fast. The first priority of survival is getting protection from the extreme weather.
>
> Bear Grylls

WEATHER PEOPLE

William 'Clement' Lindley Wragge (1852–1922): The observer who climbed a mountain

Any hobby may turn into an obsession, as it did with Luke Howard and his study of clouds. Sometimes an amateur interest becomes the spur for people to take up a particular occupation or profession. An obsession with the weather led to Clement Wragge making a career as a meteorologist.

Generally known as 'Clement', after his father, he developed a great interest in many fields of study, including astronomy, geology and meteorology, and travelled extensively overseas. Born in Stourbridge, Worcestershire, he initially studied law and became articled to a London solicitor. After a few years, deciding that the law was not for him, he resigned his articles and actually trained as a midshipman. In 1876 he moved to Australia and started work in the surveyor's department in Adelaide, South Australia, returning to England in 1878.

Wragge is probably best remembered for his having volunteered to make daily observations from the top of Ben Nevis, before the establishment of a proper observatory there. From the beginning of June to mid-October 1881, Wragge made the arduous daily ascent of the 1,344 m (4,406 ft) mountain – the highest peak in the British Isles – often in dreadful weather conditions. (His bedraggled appearance on many such occasions led to his being nicknamed 'Inclement Wragge'.) With assistants, he carried out even more extensive observations in 1882. In 1883, however, when the directors of the observatory – which had been constructed in the meantime – appointed R. T. Osmond as Superintendent, Wragge, disappointed, returned to Australia, where, in subsequent years, he was heavily involved in meteorological work (including setting up observational stations) in many parts of the continent, and extending the meteorological network to New Caledonia, Tasmania and New Zealand.

He was appointed as Government Meteorologist for Queensland in 1887, and took a particular interest in tropical cyclones, which were (and remain) a major hazard in that part of Australia. Wragge was the first to assign names to individual cyclones, initially using Greek letters, but then changing to a wide range of names (including those of unpopular politicians). After Wragge retired, the naming of cyclones or hurricanes was not reintroduced for some 60 years. Wragge was a quarrelsome individual, and annoyed other meteorologists by publishing forecasts for the whole of Australia (rather than just for Queensland) and claiming that he ran the 'Chief Weather Bureau, Brisbane'. He also managed to fall out with the Premier of Queensland, Sir Robert Philp. He spent his last years in New Zealand.

FASCINATING FACTS:
THE LONGEST TORNADO TRACK

On 21 May 1950 a tornado occurred with an extremely long track, which retains the British record to this day. It is known to have initially touched down at Little London in Buckinghamshire, and then tracked north-eastwards, remaining on the ground for no less than 107.1 km (66.5 miles) to Coveney in Cambridgeshire. It then lifted from the surface, became a funnel cloud and travelled another 52.6 km (32.7 miles) to Shipdham in Norfolk, after which it moved to the coast and out over the North Sea. With such events it is often difficult to know whether a single tornado is involved, or whether the reports are actually of a series of different funnels. In this case, however, there are numerous reports of substantial and consistent damage along the track for it to be almost certain that this was a single, long-lasting tornado, which took about four hours to cover the whole distance of about 160 km (99 miles).

Blow, winds, and crack your cheeks!
rage! blow!
Your cataracts and hurricanoes spout
Till you have drenched our
steeples, drowned the cocks!

William Shakespeare, *King Lear*

NOTABLE WEATHER EVENTS

The Bristol Channel Flood, 1607

On 30 January 1607, extremely severe flooding hit both sides of the Bristol Channel. On the Welsh side, Cardiff was particularly badly affected, and flood waters extended as far as Chepstow on the river Wye as much as 3.2 km (2 miles) above where it joins the Severn. On the Devon and Somerset side, flooding was extremely extensive, with the Somerset Levels overcome, and the water extending as far as Glastonbury Tor, 23 km (14 miles) from the Severn Estuary. Overall, it is estimated that more than 2,000 individuals were drowned, with a number of villages completely destroyed, and widespread loss of livestock wrecking the local economy. Many commemorative plaques remain showing floodwaters reaching 2 m (8 ft) above sea level, although there is one chiselled mark showing the maximum height of the wave as 7.7 m (25 ft) above current sea level.

The precise cause of this extreme event is uncertain. Various contemporary accounts relate that it was a fine morning, which would tend to argue against the generally accepted cause, a storm surge caused by the strong winds around a particularly deep depression, funnelling water into the Severn Estuary at the time of a high spring tide. A deep depression would normally be accompanied by high winds, heavy cloud and rain, not sunshine. Conditions appear to have been similar to those of the disaster caused by a storm surge that hit the east coast of England in 1953. Support for the tsunami theory comes from accounts that tell how the water receded before returning as one gigantic wave. There are indications of large boulders being transported far inland. This, and other evidence, has led to the idea that the flooding came as the result of a tsunami, possibly

caused by motion on a fault that is known to exist off Ireland. One account does link the event to a tremor felt at around the correct time. The event has been likened to the extreme tsunami that devastated countries around the Indian Ocean after the exceptionally large earthquake that occurred off Sumatra on 26 December 2004.

VOLCANOES AND WEATHER

Laki, 1783

In 1783, an eruption of the Laki fissure in Iceland began on 8 June and lasted until 4 February 1784. The fissure was 27 km (16.7 miles) long. It emitted various gases, including a vast amount (estimated at no less than 120 million tons) of the poisonous gas sulphur dioxide, most in the early stages of the eruption. When the eruption began, air circulating around an unusual high-pressure zone carried the gases in a 'dry haze' first to Norway, then across Prague, Berlin and Paris, reaching England by 23 June. Deaths (mostly among outdoor workers) in central and eastern England were approximately twice the normal number for the time of year, and total additional deaths in Britain are estimated to have reached 23,000. The 'dry haze' did not disperse until the autumn. Total deaths worldwide caused by the eruption have been estimated at as many as six million.

'The Year without a Summer', 1816

Another volcanically created disaster occurred in 1816. In 1815 there were a number of volcanic eruptions, culminating in the eruption of Mount Tambora on the island of Sumbawa in Indonesia (then the Dutch East Indies). This was the most powerful eruption of historical times (although widely unrecognised at the time, because of poor communications). It ejected so much sulphur dioxide (which combines with water to give sulphuric acid droplets) into the upper atmosphere that global temperatures were reduced by at least 0.5–1.0 degrees C. That does not seem a lot, but it was sufficient to cause crop failures and major food shortages in the northern hemisphere. Temperatures were so low that 1816 came to be called 'the Year without a Summer', 'the Poverty Year', 'the Summer that Never Was', 'the Year There Was No Summer', and 'Eighteen Hundred and Froze to Death'. There was widespread famine and there were food riots across Europe, including in Britain. It is estimated that over 65,000 people died from the effects in Britain alone. This is similar to the number of deaths in Indonesia as a direct result of the eruption, although one estimate suggests that these were about 95,000. Worldwide, the death toll is considered to be far more than the six million estimated to have occurred after the 1783 eruption of Laki.

Eyjafjallajökull, 2010

The Eyjafjallajökull volcano in Iceland is completely covered by an ice cap. It is known to have erupted in 920, 1612 and between 1821 and 1823, when it created a glacial lake and a major outburst flood (a *jökullhaup*). It then erupted three times in 2010, the first on 20 March. However, it was the eruption that started in April and continued into May that caused all the problems. On 14 April, low-level eruptions increased significantly, resulting in a major emission of volcanic ash into the atmosphere. The Met Office began to provide reports every six hours of the likely spread of the ash. The pattern of winds was such that the plume was being carried almost directly towards Britain and, in response to this, UK airspace was closed on 15 April. The effect of this was not confined to Britain, because many flights from Europe to North America cross UK airspace, so travel was severely disrupted. On the same day, observers in Shetland reported a reduction in visibility and a yellowish hue to the sky. On 17 April there was a widespread fall of ash across almost all of Britain. Although weather conditions then changed, carrying the main plume away from Britain, on 24 April flights to and from the two main airports in Iceland itself were now affected. Volcanic activity decreased the following day. Flights over Britain were permitted, provided airlines carried out risk assessments and inspected aircraft before and after each flight. On 2 May, the eruption increased in activity and by 5 May a dense plume of ash moved south towards northern and western Britain. By 9 May a dense plume was reported over the Pyrenees and above France and Italy. The volcano continued to erupt, with the first lightning in the ash column on 12 May. The ash cloud became visible over northern England on 17 May, but the eruption ceased by the end of May.

Grímsvötn, 2011

Slightly more than a year after the eruption of Eyjafjallajökull, a second Icelandic volcano, Grímsvötn, erupted on 21 May. The ash was erupted to considerable altitudes, reported as being almost 20 km (12.5 miles). The weather pattern was such that the plume began to head towards the British Isles. Almost immediately, on 22 May, flights from Europe were affected. The Met Office forecast that the plume would be over the UK during the night of 23–24 May. Ash was deposited from a dense plume on the research ship *Discovery*, in the North Atlantic between Scotland and Iceland. On 24 May, layers of ash were observed from aircraft flying to Northern Ireland and over the Manchester area, one layer being reported as 305–610 m (1,000–2,000 ft) thick. On the same day, observers reported ash fall over northern Britain, and visible deposits were found on vehicles in Orkney. Significant activity at the volcano ceased on 25 May.

FASCINATING FACTS:
TORNADO AND STORM
RESEARCH ORGANISATION

The Tornado and Storm Research Organisation (TORRO) is a British organisation founded in 1974, originally formed to undertake research into tornadoes, but later expanded to cover various forms of extreme weather, including thunderstorms, lightning and severe hail. It receives and collates reports from both amateur and professional meteorologists. Tornadoes are classified on the TORRO Scale, which is defined on the basis of wind speed, rather than on the intensity of damage, as with the Enhanced Fujita Scale (EFS) that is used in the United States. It may be considered more appropriate when there are accurate measurements of wind speed.

THE TORRO SCALE

SCALE NUMBER	WIND SPEED		NAME
	m/s−1	kph	
T0	17–24	61–86	Light
T1	25–32	90–115	Mild
T2	33–41	119–148	Moderate
T3	42–51	151–184	Strong
T4	52–61	187–220	Severe
T5	62–72	223–259	Intense
T6	73–83	263–299	Moderately devastating
T7	84–95	302–342	Strongly devastating
T8	96–107	346–385	Severely devastating
T9	108–120	389–432	Intensely devastating
T10	>121	>436	Super

Note that the scale is defined in terms of wind speeds in metres per second (m/s^{-1}), but that the speeds in kph are rounded conversions and thus appear discontinuous.

Climate is what we expect,
weather is what we get.

Mark Twain

MYTHS AND MISTAKES

'The butterfly effect'

The term 'butterfly effect' has, unfortunately, become fixed in the general public's mind to mean that tiny effects will cause major changes in the weather, more or less anywhere in the world. The term arises from misinterpretation of the title of a paper 'Predictability: Does the Flap of a Butterfly's Wings in Brazil Set off a Tornado in Texas?' given in 1972 by the mathematician Edward Lorenz. The title had actually been chosen by the organiser of the conference, not by Lorenz. His argument and conclusion was, not that small changes lead to large effects, but that the question was unanswerable. Errors in weather forecasting are inevitable and come from inadequate data coverage, unavoidable inaccuracies in measurements, incomplete knowledge of the underlying physics and because the equations used for making human or computer predictions are always approximations. The term – formally known as 'sensitive dependence on initial conditioning' – should be used to suggest that there are limits to predictability, not that such an effect exists. Indeed, it has been said that the title could well have been 'Predictability: Does the Flap of a Butterfly's Wings in Brazil Prevent a Tornado in Texas?' The fact that small initial differences may give rise to major variations in final results is the basis of chaos theory and is turned to advantage in what is known as ensemble forecasting.

Hurricanes never happen (in Britain)

The phenomena that are called hurricanes are technically known as 'tropical cyclones'. The term 'hurricane' is used for storms over the North Atlantic or eastern Pacific. They are known by other names elsewhere in the world: 'cyclones' in the Indian Ocean and 'typhoons' in the western Pacific. There are various factors that govern their formation, and which do not allow them to occur over Britain. Possibly the most significant is the fact that the sea-surface temperature must be 27°C or higher. Hurricanes gain their energy from warm waters and lose their strength when they pass over cooler waters (or travel on to land). Such sea-surface temperatures are never encountered around the British Isles. However, we can experience 'hurricane-force' winds (Beaufort Force 12, > 64 knots, > 118 kph, > 74 mph) around deep depressions. In addition, the remnants of hurricanes may sometimes combine with pre-existing depressions – which are known as 'extratropical cyclones' – causing the latter to undergo dramatic deepening.

The weather is like the government, always in the wrong.

Jerome K. Jerome,
Idle Thoughts of an Idle Fellow

FASCINATING FACTS: 'TWISTERS'

Reporters writing for newspapers, radio and television tend to use the word 'twister' for a whole range of events that involve concentrated whirlwinds. In fact, there is a whole range of different weather phenomena that tend to get lumped under this term. There are whirls – a term that is sometimes used for the whole group of rotating phenomena – devils, funnel clouds, water and landspouts, gustnadoes and true tornadoes. They form under different circumstances and by different means.

Whirls

Everyone is familiar with the way in which leaves or pieces of rubbish are lifted up in a rotating column of air caused by the wind funnelling between buildings. Similar conditions may occur when hill or mountain slopes concentrate the wind in particular locations. The whirling air may then raise water from the surface of a lake or snow from an expanse of snow-covered ground. Such phenomena are often described by the term 'devil', giving water devils and snow devils.

Devils

Although the terms 'whirls' and 'devils' tend to overlap, some of the most common devils are dust devils, which arise when heating of the ground by sunshine causes a strongly rising column of air that raises dust from the ground. The rotation is often accentuated by changes in the roughness of the surface, which may contribute significantly to the overall process. Dust devils are common in many parts of the world where the surface is dry and dusty, such as in desert and tropical regions. Occasionally, a dust devil may be strong and reach so high that it is topped by a small cumulus cloud, but generally they lose their strength as they grow taller and spray out the dust at the top of the column of rising air. (Dust devils are very common on Mars, and their tracks across the surface have been photographed by orbiting satellites. Others have been recorded as they passed by surface landers.) In more temperate climates, where dusty surfaces are less common, summer will sometimes bring hay devils, formed by the same process, in which hay or straw is lifted from the surface to form a whirling column. Generally, devils are not particularly destructive, although some damage has been reported from strong dust devils.

Funnel clouds

It is fairly common to see a grey column of cloud, looking like an elephant's trunk, reaching down from the clouds, but not touching the surface. This is a funnel cloud, known to meteorologists by the term 'tuba'. Such a column is created when there is strong convection in the cloud, which draws air up from lower layers. The reduction in pressure in the centre of the rotating column causes the air to cool and water vapour to condense, producing the column of cloud. If a funnel cloud touches down on the surface, it becomes one of the next group of whirls: a spout.

Waterspouts and landspouts

When there is strong convection in cumulonimbus clouds or the cumulus variety known as cumulus congestus, it may initially create a funnel cloud hanging down from the clouds above. If the funnel touches the surface of the sea or a lake, a waterspout is created. When the funnel first touches the surface it creates what is known as the 'dark spot'. A little later a spiral pattern becomes visible on the surface, and this is followed by a stage at which there is a distinct 'bush' of spray thrown up from the surface. Generally waterspouts decay after some 5–10 minutes, but very large ones have been known to persist for over an hour. Waterspouts are generally relatively small, ranging from a few metres to about 100 m (328 ft) in diameter.

Waterspouts often move onshore when, technically, they become 'landspouts'. The latter may be created over inland areas by exactly the same process: very strong convection in suitable clouds. However, the term 'landspout' – although technically correct – is not commonly used, and such phenomena tend to be called 'twisters' or even 'tornadoes'. Although both may cause some damage, they are much weaker than true tornadoes.

Gustnadoes

The word 'gustnado' is a colloquial term for short-lived whirls that are generated by violent winds along a gust front. They are commonly associated with hurricanes and some vigorous cumulonimbus clouds. They appear as rotating columns of dust or debris. Some definitions suggest that they do not reach the ground, but merely consist of tight, rotating columns of air that have been initiated by variations in wind speed caused by obstructions such as buildings, hills and similar features, and occasionally assisted by strong convection. They may thus be regarded as particularly violent, largely invisible, whirls.

Tornadoes

Tornadoes are in a class of their own. Although superficially similar to waterspouts and landspouts, the mechanism by which they are formed is completely different, despite the precise details of the process not being fully understood. They may form under energetic cumulonimbus clouds, but the most destructive are formed beneath supercell storms. Normally, but not always, a rotating funnel cloud is seen reaching down from the sky and finally touching the ground. Only when it does so, is the rotating column described as a tornado.

A tornado initially forms as a long, horizontal, rotating roll of air. The strong updraught beneath the parent cloud lifts the centre of the rotating column, so that it forms an inverted 'U' or 'V' shape. One of the two limbs decays, but the strong up- and downdraughts on one side of the cumulonimbus cloud or supercell reinforce the rotation of the remaining limb, giving rise to the extremely high wind-speeds within the tornado. These speeds are so high that conventional meteorological instruments are destroyed and knowledge of the speeds that occur has been gained by the use of special radar equipment. The highest recorded wind-speed for a tornado is 276 knots (512 kph), found in the devastating tornado that tracked across Oklahoma on 3 May 1999.

> When two Englishmen meet, their first talk is of the weather.
>
> Samuel Johnson, *The Idler*

WEATHER LORE

'Take heed to the weather, the wind and the sky;
If danger approaches, then cocks apace cry.'

This piece of weather lore is of doubtful validity, because there are many other factors (such as the presence of predators) that will cause cocks (and other birds) to become agitated and noisy. However, recent research has uncovered an extraordinary instance of birds 'predicting' dangerous weather. The golden-winged warblers of North America are migratory birds. They had just arrived at their breeding grounds in the Cumberland Mountains in Tennessee, when they suddenly left in advance of supercell storms that were accompanied by devastating tornadoes. (The storms spawned 84 tornadoes and resulted in 35 deaths.) The birds actually flew some 1,500 km (932 miles) in five days, apparently to avoid the severe weather. Although unconfirmed, it is suspected that they were able to detect the infrasound produced by the storms – sounds well below the threshold of human hearing.

(It may be noted that one way in which it has been suggested that the approach of tornadoes could be predicted, and used to issue warnings, is by attempting to detect the infrasound that they generate.)

> Whenever people talk to me about the weather, I always feel quite certain that they mean something else.
>
> Oscar Wilde, *The Importance of Being Earnest*

HIGHS AND LOWS

Tornadoes in Britain

Area for area, more 'tornadoes' are reported annually over the British Isles than anywhere else in the world, including the United States, which does, however, experience a greater number every year, most of which are far more destructive. It is also probably true that more reports are received from Britain simply because the country is more densely populated. Few events go unreported, even though many reports in the media are greatly exaggerated. Most events in Britain are 'landspouts', resulting from extreme convection, rather than being true 'tornadoes' produced by supercell storms and the vast column of rotating air known as a mesocyclone. Some of the events are actual tornadoes, however, such as the Birmingham tornado of 28 July 2005, which left a significant trail of damage, but which was actually described as a T1 (i.e. weak) tornado.

Lord of the winds! I feel thee nigh,
I know thy breath in the burning sky!
And I wait, with a thrill in every vein,
For the coming of the hurricane!

And lo! on the wing of the heavy gales,
Through the boundless arch
of heaven he sails;

Silent and slow, and terribly strong,
The mighty shadow is borne along,
Like the dark eternity to come;
While the world below, dismayed and dumb,
Through the calm of the thick hot atmosphere
Looks up at its gloomy folds with fear.

William Cullen Bryant, *The Hurricane*

WEATHER WORDS

Cioclón (Irish) and **Seiclon** (Welsh): Cyclone.

Corwynt (Welsh): Hurricane.

Iomghaoth or **gaoth-sgrios** (Scottish Gaelic): Tornado.

> In the eye of the tornado,
> there's no more high and low,
> no floor and sky.
>
> Francis Alÿs

FASCINATING FACTS:
A WEATHER BOMB

In early December 2014, the term 'weather bomb' was being bandied about in the media, and became the source of comment and talk of forecasters trying to dramatise the weather. But the term 'bomb' is a perfectly respectable term in meteorology and one that has been in use for many years. It describes the sudden, dramatic deepening of a depression, where the central pressure drops by at least 1 millibar per hour for a whole day (24 hours). Such bombs are usually accompanied by extremely high winds, as was the case in 2014, when northern Scotland (and the Hebrides in particular) was subjected to extreme winds and exceptionally high waves, which had been generated by the long fetch (track) of those extreme winds across the Atlantic.

There is really no such thing as bad weather, only different kinds of good weather.

John Ruskin

CHAPTER 8

FORECASTING: INSTRUMENTS AND INVENTIONS

> Meteorology itself, especially as regards English weather, is very far from having reached the phase of an exact science.
>
> Richard Inwards, *Weather Lore*

WEATHER PEOPLE

Vice-Admiral Robert FitzRoy (1805–65): 'Forecasting the weather'

Although many people complain about weather forecasts, this is usually because, despite being generally accurate nowadays, forecasting is still unable to provide highly detailed information about local weather, which always depends greatly on local conditions. But forecasts have become so common that it is hard to imagine that criticism of one's forecasts would lead to suicide. But this was indeed one factor in the death of the unfortunate Robert FitzRoy.

Best known as 'Darwin's Captain', who commanded the *Beagle* throughout the voyage of 1831–36, FitzRoy was appointed Meteorological Statist to the Board of Trade in 1854, charged with assembling weather data collected at sea. He obtained data from the captains of vessels and supplied them with suitable instruments. He devised various barometers which were installed at fishing ports so fishermen could consult them before they put to sea.

The wreck of the steam clipper *Royal Charter* off Anglesey in 1859, with the loss of about 460 lives, prompted FitzRoy to analyse the prevailing conditions and develop methods of predicting the weather. He called this 'forecasting the weather', introducing the term to the language. He established land stations that reported the weather via the telegraph. The first actual forecast appeared in *The Times* in 1860. In 1861 he introduced storm warning cones that were hoisted at ports to warn mariners of imminent gales. His *Weather Book*, of 1863, was a major advance in meteorology.

Like many meteorologists since, FitzRoy was heavily criticised when some forecasts were incorrect, and they were eventually withdrawn in 1865. Suffering deep depression, personal problems, ill-health and increasing deafness, FitzRoy committed suicide on 30 April 1865.

His great significance is recognised by meteorologists, however, and in February 2002, sea area Finisterre was renamed FitzRoy, to avoid confusion with the (smaller) sea area called Finisterre by the French and Spanish meteorological services. The Met Office's headquarters in Exeter is situated on FitzRoy Road.

Sir Francis Galton (1822–1911): Creator of the first true weather map

Although perhaps best remembered as an eminent statistician, an early geneticist and pioneer in eugenics, Sir Francis Galton made a significant contribution to meteorology in that, using observations made in October 1861, he plotted the data on a map of Britain, and in doing so created the first true weather map. To enable charts to be engraved on printing blocks he invented the device, known as a pantograph, which was then used by *The Times* for that purpose.

Galton was a firm believer in rigorous 'scientific' methods, and as such was extremely critical of the methods employed by Robert FitzRoy in 'forecasting' the weather – a term that Galton greatly disliked. He chaired a committee of enquiry for the Royal Society, which was to review the advice given by the Society in 1855, examine the work of the Meteorological Department, and report on ways in which meteorology should be carried out in the future. Overall, the tone of the report was critical of the way in which FitzRoy had carried out his work, and one result was the immediate discontinuance of the storm warnings (including the hoisting of the storm cones) that

FitzRoy had initiated. After some years, however, these were reinstated. Although the report was that of the committee as a whole, it was undoubtedly strongly influenced by Galton's views that any prediction of forthcoming weather could only be carried out through a full understanding of the basic physics underlying the phenomena involved, and not from the observation and extrapolation of actual conditions. He described these as being the 'philosophical' and the 'empirical' methods, respectively.

Lewis Fry Richardson (1881–1953): Using human 'computers' to forecast the weather

Nowadays supercomputers are used to predict the weather. Yet the original suggestion for this method of weather forecasting envisaged the use of human 'computers', all working on individual parts of the overall scheme. This idea was put forward by Lewis Fry Richardson, who was a mathematician, meteorologist and pacifist. He came to it via a study of peat.

Lewis Fry Richardson was born in Newcastle-upon-Tyne, and was a life-long member of the Society of Friends. He studied at Cambridge University and became an expert mathematician, then held a number of research and teaching posts. One of these was with National Peat Industries, where he was asked to determine how to drain peat moss, taking annual rainfall into account. The mathematical equations were not directly soluble, but he managed to develop a method of obtaining approximate solutions.

Later, Richardson realised that his methods could be applied to many practical problems, in particular to weather forecasting. Appointed as Superintendent of the Eskdalemuir Observatory (Dumfries and Galloway) and encouraged by the Director of the Meteorological Office, Sir William Napier Shaw, he further developed his ideas, and had nearly completed his book *Weather Prediction by Numerical Process*, when he resigned from the Meteorological Office in 1916 and (being a conscientious objector) joined the Friends' Ambulance Unit. He completed his work by carrying out an actual computation while serving as an ambulance driver in France. He attempted to calculate the weather six hours ahead for a single day (30 May 1910), using data obtained at 07.00 GMT. The computed result appeared to be wildly inaccurate, although an analysis undertaken many years later, using methods of smoothing the data, unknown to Richardson, showed that his techniques were potentially extremely accurate. His book was eventually published in 1922.

Richardson is known among meteorologists for his hypothetical scheme for computing weather forecasts, in which he imagined a vast sphere, representing the globe, populated by 64,000 human 'computers', all calculating specific portions of the vast overall set of equations, and passing those results to a central point from which the actual forecasts were prepared and issued.

He is also known for the little rhyme that appears in his book: 'Big whirls have little whirls that feed on their velocity, and little whirls have lesser whirls and so on to viscosity.'

Richardson's methods were impractical until the development of digital computers. He was still alive when the ENIAC computer produced the first numerical weather forecast. (It took ENIAC nearly 24 hours to produce a 24-hour forecast.)

As a mathematician he published methods that were later used in the development of the theory of fractals and in chaos theory. As a pacifist, he applied some of his mathematical techniques to the analysis of the causes of war. Abandoning meteorology in 1926, he turned his attention to the measurement of perception, such as brightness, colour and pain. Although such methods were not employed at the time, they subsequently became significant in the development of experimental psychology.

James Glaisher (1809–1903): From Irish cloud to balloons

What do you do if you are a surveyor, working in County Galway and find yourself constantly surrounded by cloud? You become a meteorologist! At least that, according to James Glaisher, was what triggered his interest in the subject.

Glaisher was born in Rotherhithe in London and developed an interest in scientific subjects, including astronomy. At the age of 20 he visited the relatively nearby Royal Observatory at Greenwich, where his brother was employed as a (human) computer. After an initial spell as a surveyor, working in Ireland, Glaisher turned to astronomical work at Cambridge University. He moved to the Royal Observatory Greenwich in December 1835 as assistant to the Astronomer Royal, G. B. Airy. When the magnetic and meteorological department was established in 1838, Glaisher was made superintendent: a post he held until he resigned in 1874. His first task was to standardise instruments and observational methods. He introduced the first regular telegraphic reports and charts of the weather, which were published from 1848 in the *Daily News*, founded by Charles Dickens.

From 1846, Glaisher started to obtain observations from some 40 stations in Britain, many of which he established, including in his duties instruction of the observers and regular inspection of the instruments.

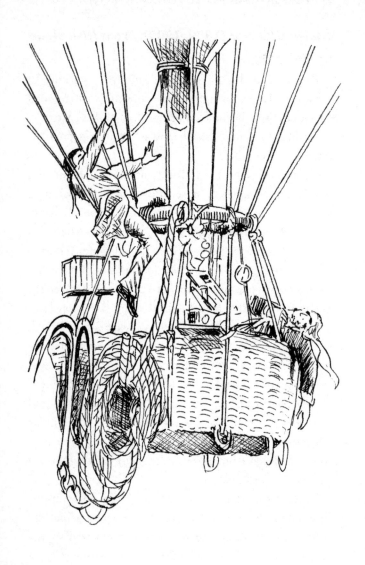

Glaisher was a member of the Royal Astronomical Society and was elected a fellow of the Royal Society in 1849. In 1850, with other members of those societies, he founded the British Meteorological Society (later the Royal Meteorological Society), of which he was secretary until 1873, except for a short break in 1867–68 when he was president.

Glaisher became famous to the general public through his ascents in balloons to obtain meteorological data at altitude. One ascent, in particular, brought him to general notice. On 5 September 1862, Glaisher ascended with Henry Coxwell, the balloonist. At an altitude that he later estimated to be 29,000 ft (approximately 8,840 m – roughly the height of Mount Everest), Glaisher became unconscious, due to (as we now realise) lack of oxygen. The balloon continued to rise and is thought to have reached as high as 37,000 ft (11,280 m). Coxwell lost the use of his hands, but was able to pull the cord to the gas-release valve with his teeth, causing the balloon to descend. Neither Glaisher nor Coxwell were permanently injured.

A more disastrous outcome accompanied a flight by the French meteorologist Gaston Tissandier. In April 1875, with two companions, he ascended to about 28,200 ft (8,600 m) at which stage he lost consciousness from lack of oxygen. When the balloon descended and Tissandier revived, he found both of his companions had died. He was permanently deafened by the experience.

Alexander Buchan (1829–1907): Buchan and his spells

Posterity sometimes plays strange tricks upon the memories of particular individuals, so that they tend to be remembered for their failures, rather than for their actual achievements. Such has been the case with the Scots meteorologist Alexander Buchan.

Buchan was educated in Edinburgh and attended Edinburgh University. In 1848 he began a career as a teacher and after a few moves became headmaster of the Free Church School in Dunblane in 1857. At this period he was particularly interested in botany and took part in an organised botanical expedition to the Alps. However, he decided that he was basically unsuited to be a teacher – partly on account of his weak voice – so decided to look for another occupation.

The Scottish Meteorological Society had been established in 1855, but initially had no less than three secretaries in its first five years. In 1860, Buchan applied for, and obtained, the position, and he retained the post until his death, 47 years later. Even whilst holding the post, as early as 1860, he found time to study at Edinburgh University and was awarded a master-of-arts degree in 1867. His indefatigable services as secretary of the Scottish Meteorological Society and his numerous meteorological publications and other work brought him national and international recognition. He was to receive dozens of awards from a whole range of international societies, including being the first recipient of the Symons Gold Medal of the Royal Meteorological Society (in 1902), and culminating in his election as a Fellow of the Royal Society in 1897. He was involved in many societies and committees, ranging from the Royal Society of Edinburgh to the London Meteorological Council, which he joined in 1887.

In 1867, Buchan suggested that a number of periods during the year were likely to exhibit unseasonal cold conditions (six periods) or warm conditions (three periods). Although based on observations for south-east Scotland, these 'Buchan spells' came to be widely applied to weather across the whole of Britain. They were often quoted in newspapers, especially in the first half of the twentieth century.

The cold periods that Buchan believed he had identified were: 7–14 February, 11–14 April, 9–14 May, 29 June–4 July, 6–11 August and 6–13 November. The warm periods were 12–15 July, 12–15 August and 3–14 December. The period in February was thought to include the coldest night of the winter, and the period in July the hottest day in the summer.

Later analysis has suggested that the spells are of limited validity and certainly cannot be used as the basis for prediction of conditions that are likely to occur. Unfortunately, his name is largely known to the general public in context of the failure of predictions based on the Buchan spells. By contrast, in the same year, 1867, Buchan published the *Handy Book of Meteorology*, which became accepted almost immediately as the standard work on the subject, used both in Britain and abroad.

In 1868 he tracked the path of a deep depression across North America and the Atlantic, and by doing so is credited with being the first to use weather charts for the purposes of weather forecasting.

Buchan was instrumental in promoting two important research organisations. The first of these was the Scottish Marine Station at Granton, under Sir John Murray, which opened in 1883. Buchan was also largely responsible for the establishment of the Observatory on Ben Nevis, also in 1883.

Between 21 December 1871 and 25 May 1876, HMS *Challenger* carried out the first global marine research expedition, achieving outstanding scientific results. (The ship's name is commemorated in the name of the very deepest part of the ocean, the Challenger Deep in the Mariana Trench, and in the names of the *Glomar Challenger* research ship and the US Space Shuttle *Challenger*.) The vast collection of oceanographical and meteorological observations was passed to Buchan, he being the person most capable of compiling comprehensive reports on the findings. This task occupied a great deal of his time, and the final *Challenger* memoirs were published in 1896.

FASCINATING FACTS:
USING WEATHER MAPS TO
FORECAST THE WEATHER

The very first weather charts showed observations obtained in 1783 from various local stations run by the Palatine Meteorological Society, in what was then the Palatinate around Mannheim (now in Baden-Württemberg) in Germany. These charts were produced in 1820 by Heinrich Wilhelm Brandes, professor of mathematics at the University of Breslau. He realised, from his charts, that storms tended to move from west to east and possibly also crossed the Atlantic before affecting Europe. Others, previously, had noted that weather systems had moved across the countryside, but their ideas had not gained general circulation or acceptance. Similarly, the work by Brandes remained largely unknown.

The suggestion that weather charts could be used to forecast the weather appears to have been first made by Urbain Le Verrier, the director of the Paris Observatory, following the destruction of 14 British and French ships by a storm in the Black Sea in 1854 during the Crimean War, but although he was responsible for collating weather observations from France, he did not follow up this idea. This was despite the fact that he had been the person to set up the first system by which meteorological observations were transmitted by telegraph to the Paris Observatory. Only when it became possible for observations from widely separated points to be sent by telegraph to a central office did it become possible to use them to draw up charts or forecast the weather. Similarly, it was also essential for the different stations to use the same standard time, so that the observations could be used to compile a picture of the weather at a specific time. In Britain, the railway network introduced the use of Greenwich Mean Time in 1847 to coordinate railway movements and this standard time was used for weather observations.

Using telegraphed observations, Vice-Admiral Robert FitzRoy drew charts showing the conditions that prevailed at the time of the loss of the *Royal Charter* steam clipper off Anglesey in 1859. These charts appear strange to modern eyes and are not particularly suitable for monitoring current weather patterns or predicting their development. The first weather chart of a form that we might recognise today was produced by Sir Francis Galton, who plotted observations for October 1861 on a map of Britain. Galton also invented a device, a pantograph, that was used to transfer charts on to the printing blocks that were used at that time. The device was used subsequently by *The Times* newspaper to produce and publish charts from information provided by the Meteorological Office. In Scotland, Alexander Buchan initiated actual forecasting using charts when he plotted the course of a depression across North America and the Atlantic in 1868, and predicted its future path.

But who wants to be foretold the weather? It is bad enough when it comes, without our having the misery of knowing about it beforehand.

Jerome K. Jerome, *Three Men in a Boat*

OBSERVATORIES AND SOCIETIES

Kew Observatory

Kew Observatory was originally founded by King George III, who was particularly interested in scientific subjects. It was built in 1769, so that the king could (and did) observe the transit of Venus – the rare passage of the planet in front of the Sun – on 3 June 1769. Meteorological observations began in 1773 (and continued until 1980). Despite its original function, relatively few astronomical observations were made in the following years, and after 1842 the observatory carried out just meteorological and magnetic observations.

In 1865, the Meteorological Committee, set up to consider the future of meteorology, recommended to the Royal Society that observatories should be established at Kew, Aberdeen, Armagh, Falmouth, Glasgow, Stonyhurst (near Clitheroe in Lancashire) and Valentia. Kew was designated as the 'Central Observatory', and was the responsibility of the British Association for the Advancement of Science. Kew came to be answerable for testing meteorological and other instruments (particularly chronometers), and supplied standard instruments to various expeditions, including one to Fort Rae in Canada as part of the (first) International Polar Year (1882–83).

In 1900, the Kew Observatory became part of the newly created National Physical Laboratory (NPL).

However, the physical premises at Kew were not large enough nor particularly suitable for NPL's purposes. In March 1902, the NPL headquarters were moved to nearby Teddington, and Kew became the 'Observatory Department', responsible for calibration, design and repair of meteorological instruments, as well as making regular meteorological observations. Because of the construction of an electric tramway near Kew, which would jeopardise magnetic measurements, a new magnetic observatory was created at Eskdalemuir in Dumfries and Galloway, which opened in 1908. (Eskdalemuir Observatory remains in operation and returns climatological and pollution data as well as magnetic and seismological records.) Control of the observatories at Kew and Eskdalemuir passed to the Meteorological Office on 1 July 1910, with many of Kew's certification functions taken over by the NPL at Teddington.

In late 1914 the suggestion was made that Kew should become the central observatory and establishment for London. A second magnetic observatory was established in 1921 at Lerwick in Shetland, Kew being unsuitable because of the general electrification of nearby railways. The bicentenary of Kew Observatory was celebrated in 1969, but by 1980, the Observatory had become outmoded, with essentially all of its former development work being carried out at the Meteorological Office's establishment at Beaufort Park, near Reading. The last meteorological observation was made on 31 December 1980. The Observatory building was then sold and is now owned by a private company.

Royal Meteorological Society

The Royal Meteorological Society is the principal society in the UK devoted to the study of the science of meteorology, in all its aspects, including weather and climate. It began as the British Meteorological Society, founded at Hartwell House, near Aylesbury in Buckinghamshire, on 3 April 1850, with ten founder members, including James Glaisher, superintendent of the magnetical and meteorological department at the Royal Observatory Greenwich. The Society held its first meeting on 7 May 1850 at which Luke Howard, among many others, joined the society. It became the Meteorological Society in 1866, when it was incorporated by Royal Charter, and was given the privilege of adding 'Royal' to its title in 1883. Membership is open to all those interested in meteorology and related subjects, such as oceanography, and includes enthusiastic amateurs as well as professional scientists.

Met Office

The initial precursor of the Met Office was established in 1854, when Vice-Admiral Robert FitzRoy was appointed Meteorological Statist to the Board of Trade as head of the meteorological department. Subsequently, it became generally known as the Meteorological Office. After World War One, in 1919, responsibility for the Office was transferred to the newly formed Air Ministry. The Office provided weather information to all the services (in addition to the general public) until 1936, when the Royal Navy set up its own meteorological service. With changes to the organisation of the armed forces, the Meteorological Office then came under the Ministry of Defence, until in 1990 it became an executive agency of the Ministry, required to act in a commercial manner. In 2000, the name was formally changed from 'Meteorological Office' to 'Met Office', as it is known to this day. In 2011, the Met Office became part of the Department for Business, Innovation and Skills.

The Observatory on Ben Nevis

It was in the 1870s that the Scottish Meteorological Society first suggested that an observatory should be built on the top of Britain's highest mountain, Ben Nevis (1,344 m/4,406 ft). There was a long delay, because of government reluctance to provide the necessary funds. In 1881 and 1882, during the summer months, Clement Wragge famously made daily ascents to carry out measurements (see page 166). Eventually, the Observatory was built, opened in 1883 and financed by the Society. The Observatory was permanently manned and conditions in the winter months were particularly arduous and it was often difficult to obtain observations because of the extremely high winds and heavy icing. The Observatory continued work until 1904, when it had to be closed because of insufficient government funding. The 20-year period during which the Observatory operated provided the longest series of observations from a mountain summit in Britain until the automatic weather station was established on the Cairngorm Summit (1,245 m/4,085 ft) in 1981.

A low-level, collaborating observatory was later constructed at Fort William, close to the foot of the mountain. Valuable comparisons thus became possible with observations from the two widely different altitudes. The low-level observatory began operating in July 1890 and the observers alternated between the two stations.

The Shipping Forecast

The Shipping Forecast broadcast by the BBC has become a much-loved British institution, and any slight changes to names, timing or other features of the broadcast are met with a barrage of messages from the general public. There were, for example, violent objections when in 2002 it was proposed to change the name of sea area Finisterre to FitzRoy. This was to avoid confusion with the smaller sea area, similarly named, used by the French and Spanish meteorological services. In that case, the change was implemented, but on other occasions public opinion has prevailed, in particular over the timing of the late-night broadcast at 00.48 GMT. Other broadcasts are made at 05.20, 12.01 and 17.54 GMT. All broadcasts are made on long-wave frequencies to ensure that they may be received anywhere around the British Isles and at considerable distances into the Atlantic. Certain broadcasts (at 00.48 and 05.20 GMT) are also carried on VHF (FM) wavelengths for reception in coastal waters. Reports from all coastal stations are broadcast at 00.48 GMT and for a selection at 05.25 as part of the broadcast at 05.20 GMT. An inshore waters forecast is given at 00.55 and at 05.27.

The Meteorological Office began issuing marine weather forecasts, transmitted by radio, in 1911. These were discontinued during and after World War One, recommencing in 1921. The shipping forecasts broadcast by the BBC began in 1924, when 14 sea areas were covered. Eight years later, the number and boundaries of the areas were modified to give just 12 individual sea areas. Following the hiatus caused by World War Two, the broadcasts resumed in 1945, but in 1949 they were expanded to give much greater coverage, with a total of 26 sea areas. Further changes were introduced in 1956, when 30 sea areas were covered, including two to the north and north-west of Iceland: North Iceland and Denmark Strait. Reports from 11 coastal stations were given. More changes came in 1984, when the two small areas of North Utsire and South Utsire were added in the North Sea. The areas beyond Iceland have now been omitted and Trafalgar added in the far south to give a total of 31 sea areas. Reports from as many as 22 coastal stations are now given in the main broadcast at 00.48 GMT. (See end of this section for sea areas, coastal stations, inshore waters and seasons.)

FASCINATING FACTS:
THE STEVENSON SCREEN

The standard enclosure for thermometers at official sites in Britain and many other countries is known as a Stevenson screen. It consists of a white-painted, box-like structure with double-louvred sides, a double roof and a floor, specifically designed to allow a free flow of air through it but also to protect the thermometers from direct exposure to sunlight. (Such enclosures have sometimes been mistaken for beehives.) The name has become associated with the eminent civil engineer, Thomas Stevenson (1818–87). He was one of the Scottish family of lighthouse engineers, the 'Lighthouse Stevensons' and he designed and constructed many famous lighthouses around Scotland. He was also the father of Robert Louis Stevenson, the author. The louvred design for a thermometer enclosure was originally proposed by others, but Stevenson first suggested the use of double louvres to prevent radiation from affecting the readings given by the thermometers. Small screens normally contain four thermometers. These are one dry-bulb and one wet-bulb thermometer, the latter's bulb being kept damp by a cover and wick dipping into distilled water. (The difference between the two temperatures registered may be used to derive the air's humidity.) The other two thermometers each contain an index to indicate the maximum and minimum temperatures attained since the last reading, and are reset each time an observation is made. Larger enclosures of similar basic construction are often used to accommodate additional instruments, such as recording thermographs or hygrographs (which provide a continuous record of temperature and humidity, respectively).

FASCINATING FACTS:
STORM CONES

Vice-Admiral FitzRoy started his storm warning service in 1861, sending messages by telegraph to various ports around the country. The following year, 1862, he introduced the use of storm-warning signals that were hoisted on masts onshore where they would be readily visible at sea. Cones and drums were used during the day and a corresponding set of lights at night. The cones and drums were made of canvas over a wooden framework and were about 1 m (3 ft) high. A cone with point upwards indicated a gale from the north, and one with point downwards, a gale from the south. A single drum indicated stormy conditions but with unknown direction. When a cone was combined with a drum it indicated a violent storm from the direction shown by the cone. The night-time signals duplicated those shown during the day. (FitzRoy even went so far as to specify the height of the mast on which the signals were to be displayed.)

Although the warnings were discontinued after FitzRoy's death, because a committee of the Royal Society reported that weather predictions were not based on sufficiently 'scientific' principles, there was considerable pressure for them to be reinstated. After an unsuccessful trial of a semaphore-type set of signals, FitzRoy's system was reintroduced in 1874, although use of a single drum was not employed. The signals were displayed in the same form until 1 June 1984, when they had been superseded by radio broadcasts and other methods of providing storm warnings.

Storm cones

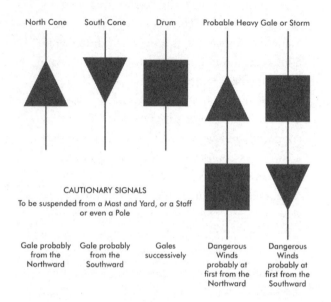

North Cone	South Cone	Drum	Probable Heavy Gale or Storm

CAUTIONARY SIGNALS
To be suspended from a Mast and Yard, or a Staff
or even a Pole

Gale probably from the Northward	Gale probably from the Southward	Gales successively	Dangerous Winds probably at first from the Northward	Dangerous Winds probably at first from the Southward

FASCINATING FACTS:
NUMERICAL WEATHER
PREDICTION

The basis for numerical weather prediction, as it is known (used worldwide for all major forecasts nowadays), is the concept of having a grid of points covering an area – in many cases, the whole Earth – at a number of levels throughout the depth of the atmosphere. Data from observations at as many grid points as possible are entered into the system, and from the known physical laws, complex mathematical computations are carried out to determine the most likely changes that will occur at every grid point at specific intervals in the future. The basic idea for this type of computation was suggested in the early twentieth century by Lewis Fry Richardson, but could only be put into practice with the introduction of digital computers. The computations are exceptionally complicated and, even for forecasts just a few days ahead, data points are required from all over the globe, hence the use of the most advanced (and expensive) supercomputers. One of the models used by the Met Office calculates data for no less than 70 levels in the atmosphere.

Spring: Slippy, drippy, nippy.
Summer: Showery, flowery, bowery.
Autumn: Hoppy, croppy, poppy.
Winter: Wheezy, sneezy, breezy.

Sydney Smith

WEATHER LORE

> 'A piece of seaweed hung up will
> become damp before it rains.'

Partly true. Seaweed does absorb atmospheric humidity and the air does become humid before it rains, but it will also become humid in fine weather sometimes, for example if dew is forming.

Weather stones

A humorous display that purports to describe the weather – and also makes fun of the complexities of scientific forecasting – is found in the 'weather stones', which are surprisingly common. They usually consist of a stone (a pebble) suspended on the end of a piece of cord, together with a note of how to interpret the 'stone'. Some examples of the instructions for 'reading' the weather stone include:

> 'If the rock is wet, it's raining.
> If the rock is swinging, the wind is blowing.
> If the rock casts a shadow, the sun is shining.
> If the rock does not cast a shadow
> and is not wet, the sky is cloudy.
> If the rock is not visible, it is foggy.

If the rock is white, it is snowing.
If the rock is coated with ice, there is a frost.
If the ice is thick, it's a heavy frost.
If the rock is bouncing, there is an earthquake.
If the rock is under water, there is a flood.
If the rock is warm, it is sunny.
If the rock is missing, there was a tornado.
If the rock is wet and swinging
violently, there is a hurricane.'

FASCINATING FACTS:
WEATHER SATELLITES

The images from weather satellites have become so common nowadays that they are found in many television weather forecasts but their concept originated when adapted forms of the German V2 rocket were used as early as 1947 to obtain photographs of clouds from high altitude. The images clearly showed the curvature of the Earth – a novelty at the time – and gave useful indications of cloud cover. The short duration of such rocket flights led to the idea of using orbiting satellites for longer-term observations. Only those with long memories can recall the first, primitive results from the TIROS-1 satellite (Television Infrared Operational Satellite), launched by NASA on 1 April 1960. Just as with early black-and-white televisions, the images consisted of multiple, clearly visible scan lines with a low resolution. However, TIROS-1 proved that observation from space would be extremely valuable in understanding and predicting the weather, and TIROS satellites began continuous coverage of the Earth in 1962. The very first global mosaic of images was constructed from TIROS data for 13 February 1965.

Modern weather satellites are extremely sophisticated and do not just return images obtained at visible wavelengths. They carry out what is termed 'topside sounding', determining various important factors, such as the amount of water vapour in the atmosphere, the temperature of the surface and cloud-tops, wind speed and direction, the presence of airborne particles and even, with the latest instrumentation, the atmospheric pressure at the surface. The data returned by meteorological and environmental satellites is used for many other purposes in addition to weather forecasting.

Polar-orbiting satellites

The TIROS satellites and the successor series, including the present-day NOAA satellites (launched by the US), the MetOp satellites (operated by EUMETSAT, the European Organisation for the Exploitation of Meteorological Satellites), the Russian Meteor and Resurs satellites, and other similar satellites, operate in what are known as 'polar orbits'. These are orbits highly inclined to the equator (close to 90°) at altitudes of 800–1,000 km. Although many satellites do not pass directly over the poles, they provide good coverage of the whole globe, including the polar regions. The orbital periods are about 90–100 minutes and the orbits may be considered to be relatively fixed in space. These orbits are Sun-synchronous, meaning that the satellites return to the same location, relative to the Sun and the Earth below, twice a day. As the Earth rotates beneath the orbits, the satellites cover adjacent swathes of the surface at each pass. The low altitude means that the instruments provide a very high resolution, and some are able to detect very low light levels so they can actually 'see' in the visible range at night.

Geostationary satellites

The second principal class of weather satellites are in geostationary orbits at an altitude of 35,900 km above the equator, at which altitude they complete one orbit in exactly the same time as one rotation of the Earth, so they appear to be stationary over a fixed point on the surface. At this altitude they are able to provide continuous coverage of the whole region visible from the satellite, but this is limited in the east–west extent and only provides adequate coverage to latitudes of about 60° North and South. Because of their far greater altitude than the polar-orbiting satellites, they are unable to provide such high-resolution imagery but, as compensation, their coverage of an area is continuous throughout day and night, rather than limited to twice a day. A series of geostationary satellites located at different longitudes provides complete coverage around the equator. Geostationary satellites include NASA's GEOS series, the Meteosat satellites, as well as Russia's Elektro-L, Japan's GMS satellites and China's Fengyun satellites. The Indian INSAT geostationary satellite also carries meteorological instrumentation.

Geostationary satellites, with their continuous coverage, are particularly valuable in tracking the development of potentially devastating tropical cyclones (hurricanes, cyclones and typhoons).

FASCINATING FACTS:
ONLY THE BEST FOR
BRITISH WEATHER

The new supercomputer for the Met Office (known as a Cray XC40) will be operational in 2017. Located at Exeter, it will weigh no less than 140 tonnes. It will have a storage capacity of 17 petabytes – a petabyte is one million gigabytes – and be able to carry out calculations at a rate of 16 petaflops – 16×10^{15} floating-point operations per second for the mathematically minded – that's 16,000 million million calculations per second, more than 100 times as fast as the existing supercomputer. The new equipment will replace the current IBM Power6 supercomputer, which operates at a 'mere' 140 teraflops – 140×10^{12} floating-point operations per second – and at present handles the Met Office's global model with a resolution of 17 km (10.5 miles) and the local grid with a resolution of 1.5 km (0.9 miles), at 70 levels in the atmosphere.

> In some high mansion, I suppose
> The weather-men confront the stars.
> Giving 'the glass' tremendous blows,
> And drinking deep at isobars.
>
> A. P. Herbert

HIGHS AND LOWS

Although the general public tends to regard forecasts as incorrect if, for example, the forecast is for 'showers', and it so happens that no rain falls at any particular spot, there are ways in which the overall accuracy of forecasts may be assessed. One method is to compare the predicted temperatures, for example, given by the early-morning forecast with those that apply later in the day. This is one way in which the Met Office determines the accuracy of its forecasts. As such:

- For lowest temperature the following night, 85.2 per cent of forecasts are within +/- 2 degrees C.

- For maximum temperature the following day, 90.6 per cent of forecasts are within +/- 2 degrees C.

- For temperatures three hours ahead, 94 per cent are within the range of +/- 2 degrees C.

- For general types of weather, such as 'rain' or 'sun', forecasts of rain three hours ahead are found to be 73.3 per cent accurate. However, as already mentioned, rainfall may be so localised that only the overall accuracy may be established.

- For forecasts of sunshine three hours ahead, 78.4 per cent of forecasts are correct.

Overall, the Met Office has found that current three-day forecasts are as accurate as those made in 1980 for just a single day ahead.

FASCINATING FACTS:
LUSTRUM

This is a term rarely encountered under normal circumstances, but it is used quite frequently in meteorology and climatology. It means a period of five years, and is often used when describing climatic conditions.

FASCINATING FACTS:
CHAOS THEORY

Chaos theory is a branch of mathematics that deals with the behaviour of mechanical or physical systems (weather being of particular interest to us). These systems have been found to behave unpredictably over time, despite being apparently governed by simple, well-known laws. Any slight change in the initial conditions in such systems may lead to very different results at a later stage. In weather forecasting this variation is actually put to good use in what is known as 'ensemble forecasting'.

FASCINATING FACTS:
ENSEMBLE FORECASTING

No matter how well the measurements are made, nor how well-established are the equations that describe how conditions change over time, and even with the use of gigantic, expensive supercomputers, chaos theory tells us that there are always unavoidable errors in any predictions. However, this is put to good use in ensemble forecasting. The calculations are run a number of times with very slightly different initial conditions (pressure a little bit higher here, wind speed a bit different there, etc.). If at the end of a run the results of several calculations are very similar, then one can have confidence that the forecast is basically sound. If, however, the results differ widely, then one is warned that the predictions may be unreliable. In this way, forecasters have an indication of the probable trustworthiness of their forecasts.

Though the Clerk of the Weather insist,
And lay down the weather-law,
Pintado and gannet they wist
That the winds blow whither they list
In tempest or flaw.

Herman Melville, *Pebbles*

WEATHER WORDS

Rhagolygon y tywydd (Welsh), **Réamhaisnéis na haimsire** (Irish) and **Wather forecast** (Scots): Weather forecast.

Tywydd (Welsh), **Aimsir** (Irish) and **Wather** (Scots): Weather.

Origin of the word 'weather'

The word 'weather' stems from Old English *weder*; related to Old Saxon *wedar*, Old High German *wetar* and Old Norse *vethr*.

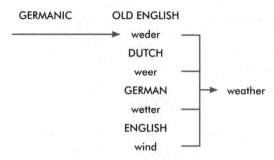

Old English weder, of Germanic origin; related to Dutch weer
and German wetter, probably also to the noun wind.

FASCINATING FACTS:
SEA AREAS

There are 31 sea areas around the British Isles that are used for the Shipping Forecast broadcast by the BBC. These are given in a fixed, specific order, running clockwise around Britain. First areas in the North Sea, followed by the English Channel, Western Approaches, Irish Sea and, finally, the North Atlantic. One area, the southernmost, Trafalgar, is mentioned only in the broadcast at 00.48 GMT, unless there is a gale warning for the area, in which case details are always included.

Viking	Humber	Lundy
North Utsire	Thames	Fastnet
South Utsire	Dover	Irish Sea
Forties	Wight	Shannon
Cromarty	Portland	Rockall
Forth	Plymouth	Malin
Tyne	Biscay	Hebrides
Dogger	Trafalgar	Bailey
Fisher	FitzRoy	Fair Isle
German Bight	Sole	Faeroes
		Southeast Iceland

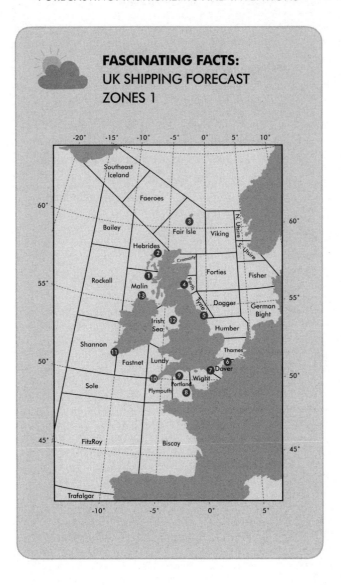

FASCINATING FACTS:
UK SHIPPING FORECAST
ZONES 1

FASCINATING FACTS:
COASTAL STATIONS

The main broadcast of conditions forecast for the sea areas is followed by reports from various coastal stations. Here, all individual station reports are given in the main broadcast at 00.48 GMT, but some are omitted from later broadcasts. Again, they are given in a clockwise sequence.

Tiree Automatic (1)
Stornoway (2)
Lerwick (3)
Wick Automatic (00.48 broadcast only)
Aberdeen (00.48 broadcast only)
Leuchars (4)
Boulmer (00.48 broadcast only)
Bridlington (5)
Sandettie Light Vessel Automatic (6)
Greenwich Light Vessel Automatic (7)
St. Catherine's Point Automatic (00.48
  broadcast only)
Jersey (8)
Channel Light Vessel Automatic (9)
Scilly Automatic (10)
Milford Haven (00.48 broadcast only)

Aberporth (00.48 broadcast only)
Valley (00.48 broadcast only)
Liverpool Crosby (00.48 broadcast only)
Valentia (11)
Ronaldsway (12)
Malin Head (13)
Machrihanish Automatic (00.48 broadcast only)

FASCINATING FACTS:
INSHORE WATERS

The forecasts for inshore waters follow the same clockwise pattern as that for the main sea areas and coastal stations. With a few exceptions, most of the areas are taken from one prominent headland to another.

1. Cape Wrath – Rattray Head, including Orkney
2. Rattray Head – Berwick-upon-Tweed
3. Berwick-upon-Tweed – Whitby
4. Whitby – Gibraltar Point
5. Gibraltar Point – North Foreland
6. North Foreland – Selsey Bill

7. Selsey Bill – Lyme Regis
8. Lyme Regis – Land's End, including the Isles of Scilly
9. Land's End – St David's Head, including the Bristol Channel
10. St David's Head – Great Orme's Head, including St George's Channel
11. Great Orme's Head – Mull of Galloway
12. The Isle of Man
13. Lough Foyle – Carlingford Lough
14. Mull of Galloway – Mull of Kintyre, including the Firth of Clyde and the North Channel
15. Mull of Kintyre – Ardnamurchan Point
16. The Minch
17. Ardnamurchan Point – Cape Wrath
18. Shetland Isles
19. The Channel Islands

Inshore waters

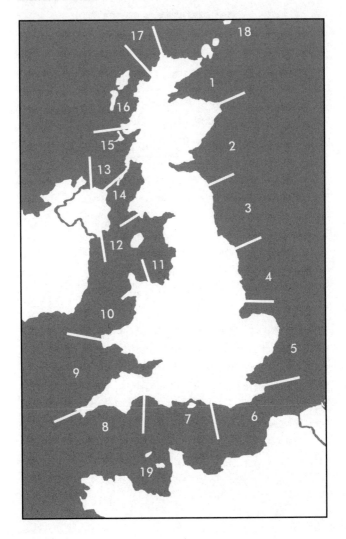

FASCINATING FACTS:
THE SEASONS

In general usage, the seasons in countries (such as in Britain) in the Earth's temperate zones are regarded as those that have an astronomical basis: more specifically, on the position of the Earth in its orbit around the Sun. The seasons are taken to begin shortly before the equinoxes (March and September) and solstices (June and December), so that we have:

Winter	December to February
Spring	March to May
Summer	June to August
Autumn	September to November

In Britain, however, there are effectively five distinct seasons, if one divides the year on the basis of meteorological conditions, particularly temperature and overall weather patterns.

Early winter	late November to late January
Late winter	late January to end of March
Spring and early summer	April to mid-June
High summer	mid-June to early September
Autumn	early September to late November

Elsewhere in the world, the seasons naturally differ considerably. In the tropics, temperatures are generally high throughout the year, so the division is usually into dry and rainy seasons, or else is referred to the prevailing monsoon winds. In subtropical regions, where there are greater temperature variations, the terms 'hot season' and 'cold (or cool) season' tend to be used instead. At high latitudes, there is often an abrupt change between winter and summer (or summer and winter) conditions, so spring and autumn may be extremely short or essentially non-existent.

Turn your face to the sun and the shadows fall behind you.

Maori proverb

FOR THE LOVE OF
RADIO 4

AN UNOFFICIAL
COMPANION

CAROLINE HODGSON

FOR THE LOVE OF RADIO 4
An Unofficial Companion

Caroline Hodgson

ISBN: 978-1-84953-642-4   Hardback   £9.99

'If you love Radio 4 it's impossible to turn it off. If you read this book it's impossible to put down.'
Charles Collingwood

From *Farming Today* at sunrise to the gentle strains of 'Sailing By' and the *Shipping Forecast* long after midnight, Radio 4 provides the soundtrack to life for millions of Britons. In *For the Love of Radio 4*, Caroline Hodgson celebrates all that's best about the nation's favourite spoken-word station, taking us on a tour through its history, its key personalities and programmes, and countless memorable moments from the archives.

'I found the book to be full of fascinating detail. It is clearly a labour of love, perfectly designed for Radio 4 lovers.'
Simon Brett

Have you enjoyed this book?
If so, why not write a review on your favourite website?

If you're interested in finding out more about our
books, find us on Facebook at **Summersdale Publishers**
and follow us on Twitter at **@Summersdale.**

Thanks very much for buying this Summersdale book.

www.summersdale.com